Disaster Risk Reduction for the Built Environment

Disaster Risk Reduction for the Built Environment

Lee Bosher
Loughborough University, UK

Ksenia Chmutina
Loughborough University, UK

WILEY Blackwell

Registered Offices
John Wiley & Sons, Inc., 111 River Street, Hoboken, NJ 07030, USA
John Wiley & Sons Ltd, The Atrium, Southern Gate, Chichester, West Sussex, PO19 8SQ, UK

Editorial Office
9600 Garsington Road, Oxford, OX4 2DQ, UK

For details of our global editorial offices, customer services, and more information about Wiley products visit us at www.wiley.com.

Wiley also publishes its books in a variety of electronic formats and by print-on-demand. Some content that appears in standard print versions of this book may not be available in other formats.

Limit of Liability/Disclaimer of Warranty

Library of Congress Cataloging-in-Publication Data Applied For

ISBN: 9781118921494

Cover image: Detail of a 1:50 scale-model of the city of Jerusalem during the late Second Temple Period. Photograph taken by Lee Bosher in January 2012 at the Israel Museum in Jerusalem.

Cover design by Wiley

Set in 10/12pt Warnock by SPi Global, Chennai, India
Printed and bound in Singapore by Markono Print Media Pte Ltd

10 9 8 7 6 5 4 3 2 1

Contents

List of Figures

Figure	Page

List of Tables

Note on the Authors:

Dr Lee Bosher is a Senior Lecturer in Disaster Risk Reduction in the Water, Engineering and Development Centre (WEDC) at Loughborough University, England. He has a background in disaster risk management and his research and teaching includes disaster risk reduction and the multidisciplinary integration of proactive hazard mitigation strategies into the decision-making processes of key stakeholders, involved with the planning, design, construction and operation of the built environment. Lee is coordinator of the *International Council for Building's* Working Commission W120 on 'Disasters and the Built Environment', a Fellow of the *Royal Geographical Society* and he has been involved in research projects that investigated how urban resilience can be increased in the UK, Haiti, India, Nigeria and across parts of Europe. Lee's previous books include *'Hazards and the Built Environment'* (2008) and *'Social and Institutional Elements of Disaster Vulnerability'* (2007).

Dr Ksenia Chmutina is a Lecturer in sustainable and resilient urbanism in the School of Civil and Building Engineering, Loughborough University. Her main research interest is in synergies of resilience and sustainability in the built environment, including holistic approaches to enhancing resilience to natural hazards and human-induced threats, and a better understanding of the systemic implications of sustainability and resilience under the pressures of urbanisation and climate change. She has extensive experience of working on RCUK and EU-funded projects that have focused on resilience and sustainability of urban spaces in Europe, China and the Caribbean.

Also many thanks to **Dr Alister Smith** for authoring the important chapter on Landslides (Chapter 7). Alister is a Lecturer in Infrastructure in the School of Civil and Building Engineering at Loughborough University. He is a Civil Engineer specialising in Geotechnical Engineering and Intelligent Infrastructure.

Foreword

Disasters are an existential threat to the long-term sustainable development of humanity on this planet. Between 2010 and 2015 the world experienced 530 disaster events that affected 140 million people, killed 78 thousand people and caused US$151bn in damages; figures that are testament to the massive (and increasingly) negative impacts of disasters globally.

During the last few decades, documented increases of disastrous events have combined with theoretical developments that have required a fresh approach to the way in which disasters are managed. Emphasis has moved away from disaster relief and emergency preparedness, towards a more sustainable approach incorporating hazard mitigation and effective risk management. Central to this is the need for developmental practises to be more sensitive so that the impacts of a wide range of hazards and threats can be mitigated. This needs to be achieved through proactive measures. These proactive measures are likely to have a bearing on the professional training (formal and informal) and day-to-day activities of a vast range of construction practitioners and other key stakeholders; these broad 'activities' are the central focus of this book.

This textbook provides a multi-facetted introduction to how a wide range of risk reduction options can be mainstreamed into formal and informal construction decision making processes, so that Disaster Risk Reduction (DRR) becomes a core component of what could be termed the 'developmental DNA'. The contents highlight the positive roles that practitioners such as civil and structural engineers, urban planners and designers, and architects (to name just a few) can undertake to ensure that disaster risk is addressed when (re)developing the built environment. Risk management principles will be presented and illustrated with examples in the context of a range of the most prominent natural hazards in two sections focused on a) Hydro-meteorological hazards (floods, hurricanes, tornadoes) and b) Geological hazards (earthquakes, landslides and tsunamis).

The book does not set out prescriptive ('context blind') solutions to complex problems because such solutions invariably generate new problems. Instead this book raises awareness, and in doing so, the intention is to inspire a broad range of people to consider DRR in their work or everyday practices. This highly illustrated text book provides an interesting range of examples, case studies and thinking points that will help the reader to consider how DRR approaches might be adapted for differing contexts. Ultimately, it is hoped that the contents of the book will convince an expansive range of construction practitioners to incorporate DRR thinking and innovations into their everyday practice.

Acknowledgements

As one might expect from the multi-disciplinary nature of the subject matter, the authors wish to thank the myriad academics, practitioners and members of the public that have inspired us to write this textbook. We are particularly grateful to the artists, photographers, businesses and governmental and non-governmental institutions that have kindly granted us permission to use photographs and other images in this publication.

This textbook has been a labour of love for both authors but it has involved numerous long days in the office and spending far too much time away from family members. Therefore we are eternally grateful for all the support, patience and humour that our families have given us during the last two years.

The book, and the sentiments contained within, are dedicated to all people globally that strive in the face of everyday hardships and inequalities to exist, and endeavour not to be the victims of future disasters. Thus as a small token of support, any royalties obtained from this book will be donated to the Water, Engineering and Development Centre (WEDC) at Loughborough University. WEDC has been chosen because it is committed to the provision of effective, evidence-based and appropriate solutions for the improvement of basic infrastructure and essential services for people living in low- and middle-income countries. These are the critical services that provide the essential foundations for a decent life as well as for effective grass roots based disaster risk reduction.

List of Acronyms

AISC	American Institute of Steel Construction (USA)
ASCE	American Society of Civil Engineers
ASEE	American Society for Engineering Education
ASSE	American Society of Safety Engineers
BGS	British Geological Survey
BRE	Building Research Establishment
BREEAM	Building Research Establishment's Environmental Assessment Method
CABE	Chartered Association of Building Engineers
CARRI	Community and Regional Resilience Institute
CIAT	Chartered Institute of Architectural Technology
CIBSE	Chartered Institute of Building Service Engineers
CIHT	Chartered Institution of Highways and Transportation
CIOB	Chartered Institute of Building
CIRIA	Construction Industry Research and Information Association (UK)
CIWEM	Chartered Institution of Water and Environmental Management
CROSS	Confidential Reporting on Structural Safety
DEM	Department of Emergency Management (Barbados)
DRM	Disaster Risk Management
DRR	Disaster Risk Reduction
DTM	Digital Terrain Model
EA	Environment Agency (UK)
EF	Enhanced Fujita (scale)
EM-DAT	International Disaster Database
ENAEE	European Network for Accreditation of Engineering Education
EWS	Early Warning System
FEMA	Federal Emergency Management Agency
FLAG	Flood Liaison and Advice Group (UK)
GAR	Global Assessment Report (on DRR)
GHG	Green House Gas

GIS	Geographical Information System
GSJ	Geological Survey of Japan
HFA	Hyogo Framework for Action (2005-2015)
HSE	Health & Safety Executive (UK)
ICE	Institution of Civil Engineers
IDNDR	International Decade for Natural Disaster Reduction
IHE	Institute of Highway Engineers
IPCC	Intergovernmental Panel on Climate Change
IPENZ	Institution of Professional Engineers New Zealand
ISDR	International Strategy for Disaster Reduction
ISO	International Organization for Standardization
IStructE	Institution of Structural Engineers
JMA	Japan Meteorological Agency
JNURM	Jawaharlal Nehru Urban Renewal Mission (India)
JSCE	Japan Society of Civil Engineers
LEED	Leadership in Energy and Environmental Design
LRF	Local Resilience Forum (UK)
MMS	Moment Magnitude Scale
Mw	Moment Magnitude Scale
NGO	Non-Governmental Organisation
NHC	National Hurricane Centre (NOAA)
NHS	National Health Service (UK)
NIMTOO	Not In My Term of Office
NN	Normal Null (sea level)
NOAA	National Oceanic and Atmospheric Administration (USA)
PDCs	Pyroclastic Density Currents
PIRA	Provisional Irish Republican Army
RIBA	Royal Institute of British Architects
RICS	Royal Institution of Chartered Surveyors
RTPI	Royal Town Planning Institute
SCOSS	Standing Committee on Structural Safety
SDGs	Sustainable Development Goals
SFA	Sendai Framework for Action (2015-2030)
SME	Small to Medium Sized Enterprise
SUDS	Sustainable urban drainage systems
UCLG	United Cities and Local Governments
UKSPEC	UK Standard for Professional Engineering Competence
UNDP	United Nations Development Programme

UNDRO	United Nations Disaster Relief Office
UNEP	UN Environmental Programme
UNISDR	United Nations Office for Disaster Risk Reduction
USGS	U.S. Geological Survey
VEI	Volcanic Explosivity Index

List of Case Studies

List of Thinking Points

Section I

Introduction to Book and Concepts

1

Introduction

Recent natural and human-induced events have highlighted the fragility and vulnerability of the built environment to disasters. People's homes and work places, transport networks, water supply and sanitation systems, energy generation and distribution networks are all critical to the way we live our lives but they can also be vulnerable to a broad range of risks and threats. These physical systems, which represent a major proportion of the long-term developmental investment for most countries, have traditionally been designed, built and maintained by the myriad professions involved with the construction industry. Therefore, it is argued that a wide range of built environment practitioners can play a proactive role in ensuring that the built environment is designed and developed in a way that alleviates or eradicates the current and future risks that natural hazards and human-induced threats pose.

> *"While hazards, such as earthquakes, cyclones and tsunamis are natural in origin; the way that disaster risk has become embedded in the contemporary urban landscape is largely anthropogenic. Decades of mass urbanisation accompanied by poor urban planning, non-existent or poorly regulated building codes and little or no proactive adaptation to the impacts of climate change have increased humanity's exposure to these hazards".* (Bosher 2014: 240)

Clearly there are many deeply ingrained root causes that can lead to some people being poor/ marginalised or particularly prone to the impacts of natural hazards; these root causes are especially focused upon in publications such as '*At Risk*' by Wisner *et al.* (2004) and thus are not discussed in as much depth in this book. '*Disaster Risk Reduction for the Built Environment: An introduction*' examines multi-hazard and multi-threat adaptation issues mainly from a built environment perspective whilst also considering the non-structural elements of multi-hazard/ threat adaptation.

The aim of this book is to highlight the numerous positive roles that practitioners such as civil and structural engineers, urban planners and designers, architects and facilities managers (to name just a few) can undertake to ensure that disaster risk is attended to in on-going and future construction projects. The book does not aim to set out prescriptive ('context blind') solutions to complex problems because such solutions can invariably generate new problems. It is intended that this book can raise awareness, and in doing so inspire a broad range of people (professional or lay) to consider Disaster Risk Reduction (DRR) in their work or everyday activities. We will achieve this by providing a broad range of examples, case studies and thinking points that will help to facilitate consideration of differing contexts.

Disaster Risk Reduction for the Built Environment, First Edition. Lee Bosher and Ksenia Chmutina.
© 2017 John Wiley & Sons Ltd. Published 2017 by John Wiley & Sons Ltd.

Figure 1.1 Locals dealing with the aftermath of the 2015 Nepalese earthquake (*Source:* US Embassy Kathmandu on Flickr).

1.1 So what is a Disaster?

Disaster is a serious disruption of the functioning of a society, causing widespread human, material, or environmental losses which exceed the ability of the affected society to cope using its own resources.

Alexander (2002) outlines some of the characteristics of disasters, including 'substantial' destruction and/or 'mass' casualties, without putting values on the scale of the disaster as "*small monetary losses can lead to major suffering and hardship or, conversely, large losses can be fairly sustainable...*", depending on the chain of circumstances.

Wisner *et al.* (2004) state "*A disaster occurs when a significant number of vulnerable people experience a hazard and suffer severe damage and/or disruption of their livelihood system in such a way that recovery is unlikely without external aid. By 'recovery' we mean the psychological and physical recovery of victims, the replacement of physical resources and the social relations required to use them.*"

Key points arising from these definitions are the requirement for external aid resulting from a combination of an event and people's abilities (or inabilities) to cope with the event. The disaster event is what leads to the emergency response to meet the needs of the affected population.

1.2 What are the Hazards and Threats?

A '**hazard**' can be defined as a dangerous phenomenon, substance, human activity or condition that may cause loss of life, injury or other health impacts, property damage, loss of livelihoods and services, social and economic disruption, or environmental damage.

A 'threat' can be defined as a person or thing that is regarded as dangerous or likely to inflict pain or misery.

For the purposes of this book, the definitions that will be used to distinguish between the two main causes of disasters are the simpler descriptors of:

Hazard: is primarily a 'natural' source of potential danger
Threat: is primarily a 'human-induced' source of potential danger

Disasters are usually classified into natural and human induced (sometimes also called 'manmade'). 'Natural disasters' is a common term used, particularly by the media, as it relates to disasters that appear to have been caused by hazards of natural origin such as extreme weather, geophysical phenomena or epidemics. However, it is important to recognise that these so called 'natural disasters' are rarely very natural, because there tends to be many important human induced factors that have converted the natural hazard into a disaster (i.e., low-quality buildings, poor locational planning), see Thinking Point *1.1* and later chapters for further discussions on this matter.

Natural hazards are typically split into two categories, namely, 1) geo-hazards, and 2) hydro-meteorological hazards. Table 1.1 provides a list of the key geophysical-hazards and hydro-meteorological hazards that occur globally. The magnitude of natural hazards tends to be determined by key factors such as meteorology (which is influenced by the changing seasons), topography, hydrology, geology, biodiversity (of flora and fauna) and tidal variations (caused by lunar and meteorological influences, coastal topography and influenced by the type and locality of coastal developments). These processes are typically benign and provide the basis for people to exist in harmony with their natural environment. However, infrequently (and some would suggest more frequently) natural hazards impact upon the built environment, causing damage, deaths, disruption and financial losses.

1.3 Climate Change and Disasters

There is now a broad scientific consensus (including from the Intergovernmental Panel on Climate Change, IPCC) that the global climate is changing in ways that are likely to have a profound impact on human society and the natural environment over the coming decades (IPCC, 2014). Experts have suggested that the impact of global climate change (which is arguably both natural and anthropogenic in nature) has increased the frequency and intensity of disasters, and will further increase the frequency and intensity of such events in the future (IPCC, 2014). The impact of these events can be psychological, sociological and political but are typically reported in economic terms.

Thinking Point 1.1

So How Natural are 'Natural Disasters'?

So why is a 'natural disaster' not really a 'natural disaster'? Can you think of any examples when it would be clearly appropriate to call a disaster a 'natural disaster'? The human influences upon the causes of disasters are too often overlooked because sometimes these influences can be discrete and driven by very different socio-economic factors. For example, in many high-income countries, people like to live near rivers (and are prepared to pay for the benefit in many cases) for the aesthetic and recreational benefits that rivers can offer. Therefore, a flood event that occurs in the non-tidal stretch of the River Thames, for example, inundating people's homes, businesses and lifelines will typically be referred to as a 'natural disaster' but the flood hazard manifests itself as a disaster, because, in this case, society has chosen to build homes, infrastructure and businesses in an area vulnerable to floods.

Table 1.1 Typology of Hazards and Threats.[1]

Natural hazards A Hazard is primarily a 'natural' source of potential danger		Human-induced threats A 'Threat' is primarily a 'human-induced' source of potential danger	
Geophysical Hazards	Earthquakes	*Malicious*	War
	Volcanic eruptions		Terrorism
	Tsunamis (*inc. Seiches*)		Arson
	Landslides		Civil unrest
	Subsidence		Vandalism
Hydro-Meteorological Hazards	Floods	*Non-malicious*	Ineffective planning
	Coastal erosion		Poor quality construction
	Hurricanes/cyclones/typhoons		Rapid urbanisation
	Tornadoes		Pollution
	Extreme temperatures		Epidemics
	Storm surges		Industrial 'accidents'
	Drought		Corruption
	Fires		

Sea-level rise could provide an increasing challenge to developments located in coastal areas. Inevitably, such changes will have, and are already beginning to have, major consequences for the built environment, particularly critical infrastructures. The potential interrelationships between climate change, anthropogenic systems and natural systems, and their subsequent influences on natural hazards and human-induced threats are illustrated in Figure 1.2.

Changes in many extreme weather and climate-related events have been observed since the mid-twentieth century. Global climate change is understood to be the result of human activities since the Industrial Revolution — such as the burning of fossil fuels and land use change (e.g., deforestation) — resulting in a significant increase in greenhouse gases such as carbon dioxide (CO_2) (Figure 1.3).

While greenhouse gases are a natural part of the Earth's atmosphere and serve to maintain temperatures to support life, excessive emission of these gases is causing more heat to be trapped in the atmosphere, leading to rising temperatures. Projected changes in the climate include temperature increases on land and at sea, leading to melting of glaciers and ice caps, sea-level rise, and changing/irregular rainfall patterns. These changes affect almost every aspect of human life and the ecosystems on which it depends. Climate change will result in increases in the frequency and intensity of extreme weather events, as well as significant impacts from more gradual changes. The nature, extent and duration of climate change impacts will vary across different regions.

---

[1] Please note: An in-depth analysis of each of these hazards/threats is beyond the scope of this book. Some key hazards/threats will be discussed in sufficient depth in forthcoming chapters. For more detailed information about different types of hazards and some key threats you may wish to refer to the book 'Environmental Hazards' by Keith Smith (2013).

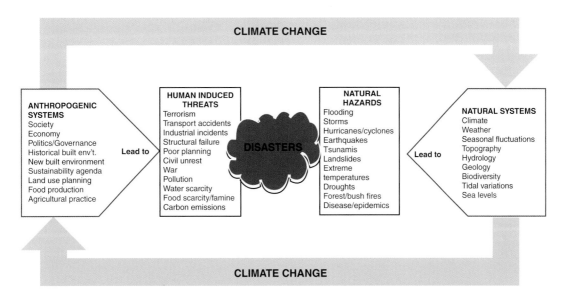

Figure 1.2 Potential interrelationships between climate change and hazards/threats (*Source:* Adapted from Bosher, 2007.).

According to the fifth IPCC assessment report, the number of cold days and nights has decreased and the number of warm days and nights has increased on the global scale, with the frequency of heat waves on the increase in large parts of Europe, Asia and Australia. There are now more land regions where the number of heavy precipitation events has increased than where it has decreased: this implies greater risks of flooding.

Climate change may not be significantly responsible for the recent skyrocketing cost of disasters, but it is very likely that the portended changes in the climate will impact future catastrophes. Climate models provide a glimpse of the future, and while they do not agree on all of the details, most models predict a few general trends. First, according to the IPCC, an increase of greenhouse gases in the atmosphere will probably boost temperatures over most land surfaces, although the exact change will vary regionally. More uncertain – but possible – outcomes of an increase in global temperatures include increased risk of drought and increased intensity of storms, including tropical cyclones with higher wind speeds, a wetter Asian monsoon, and, possibly, more intense mid-latitude storms. The combined result of increased temperatures over land, decreased equator-versus-pole temperature differences, and increased humidity could be increasingly intense cycles of droughts and floods, as more of a region's precipitation falls in a single large storm rather than a series of smaller precipitation events. A warmer, wetter atmosphere could also affect hurricanes, but changes to tropical storms are harder to predict and track. Even if tropical storms don't change significantly, other environmental changes brought on by global warming could make the storms more deadly. Melting glaciers and ice caps will likely cause sea levels to rise, which would make flooding more severe when storms affect coastlines (Figure 1.4).

Limiting climate change would require substantial and sustained reductions in greenhouse gas emissions (known as climate change mitigation), which, together with adaptation can limit climate change risks. Thus, the UN's 'Global Assessment Report on Disaster Risk Reduction' (UNISDR, 2015a) notes that the benefits of strong and early action far outweigh the economic costs of inaction. The implications

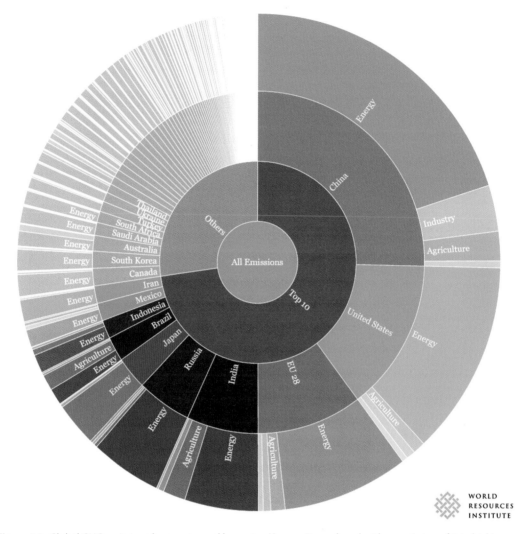

Figure 1.3 Global GHG emissions by country and by sector (*Source:* Reproduced with permission of Friedrich).

of any 'strong and early action' will pose important questions for the planning, design, construction, and maintenance of the built environment and the protection of critical infrastructures.

Consequently, when designing buildings, infrastructure and communities, it is important to plan for future climatic conditions throughout the design life of the development, and not just for the current climate (this issue will be discussed further in Chapter 9). This should be seen as a commercial opportunity for practitioners involved with the built environment. Well-designed buildings, appropriately protected from the hazards associated with climate change, will be easier to sell or let and could also command higher prices. Opportunities are therefore available for organisations to position themselves as market leaders in the concept of 'future-proofing' buildings, thereby presenting a means of attracting new customers and potentially gaining a competitive edge.

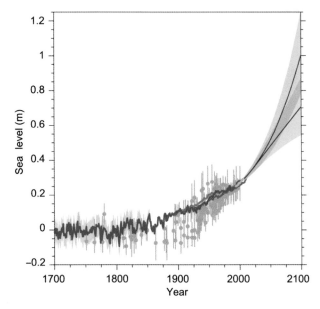

Figure 1.4 IPCC Sea level rise projections: Compilation of paleo sea level data, tide gauge data, altimeter data, and central estimates and likely ranges for projections of global mean sea level rise for RCP2.6 (blue) and RCP8.5 (red) scenarios (Section 13.5.1), all relative to pre-industrial values (2013) (*Source:* Church 2013. Reproduced with permission of Cambridge University Press).

1.4 Impacts of Disasters Globally

Over the past 40 years there has been a significant increase in the number of people affected by disasters globally. According to the Red Cross, an average of 354 disasters occurred throughout the world each year from 1991 to 1999. Between 2000 and 2004, this figure more than doubled to an average of 728 disasters per year (cited in Penuel & Statler, 2011: 233). And each year, the amount of people being affected by disasters has gradually been increasing (Figure 1.5).

Between 2000 and 2013, on average 226 million people per year (that is approximately 3% of the world's population) were affected by all types of disasters. During the same period, nearly 2 million people were killed, an average of 148,894 people each year (EM-DAT, 2014). Based on recent figures obtained from the EM-DAT disaster database (Table 1.2), over the last five years there has been a general decrease in the amount of disaster events occurring and a trend towards lower total deaths each year. 2010 experienced the most disaster events ever recorded (670) and thus witnessed a high amount of deaths (336,743), largely linked to the impacts of the Haitian Earthquake (for more details, see Case Study 5.2).

The least amount of disaster events occurring over the same period happened in 2015 (251), and as a consequence the annual economic damages were the lowest (US$14bn) over that 5-year period. The most costly year (in economic terms) was 2011, with US$364bn of damages, mainly linked to the impacts of the Japan (Tōhoku) earthquake and tsunami, Hurricane Irene (USA) and floods in Thailand, the Philippines and Pakistan. The average figures per year over this five-year period of, 530 disaster events, 140 million people affected, 78 thousand killed and US$151bn in damages are testament to the massive negative impacts that disasters have across the world.

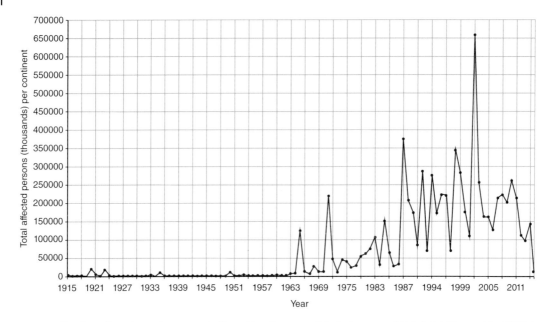

Figure 1.5 Total number of people reported affected by disasters, globally between 1915–2015 (*Source:* The OFDA/CRED EM-DAT International Disaster Database).

Table 1.2 Global Disaster Events and Impacts Between 2010 and 2015.

Year	Amount of disaster events	Total deaths	Total people affected	Total damages (US$)
2010	670	336,743	260.5m	153 bn
2011	605	40,769	213m	364 bn
2012	560	17,578	113m	156 bn
2013	546	28,922	97m	120 bn
2014	549	25,790	142m	98 bn
2015	251	19,321	11.5m	14 bn
Ave:	*530*	*78,187*	*140m*	*151 bn*

Source: The OFDA/CRED EM-DAT International Disaster Database

Disasters wreak havoc on nations irrespective of a country's wealth or resources, but invariably it is the least-developed nations that suffer the most (Bosher, 2008; UNISDR, 2015a). Amongst the top 10 countries[2] in terms of disaster mortality in 2012, six countries were classified as low-income or lower-middle income economies and four as high-income or upper-middle income economies. These countries accounted for 68% of global reported disaster mortality in 2012 (Guha-Sapir, Hoyois & Below, 2013).

---

[2] The countries listed in the report, based on total deaths in 2012, were 1) Philippines, 2) China P.R., 3) Pakistan, 4) India, 5) Russia, 6) Afghanistan, 7) Nigeria, 8) Peru, 9) Iran, and 10) USA.

1.5 Trends in the Occurrence of Disasters

When investigating disasters, it is interesting to assess whether there are any patterns or trends in the occurrence of disasters. Looking back over the last decade, the main pattern is one of major disasters dominating the death rates. Notable disasters (with approximate death tolls) include:

- 2004, earthquake and tsunami, Indian Ocean, 280,000 deaths
- 2010, earthquake, Haiti, 160,000 deaths
- 2008, cyclone Nargis, Myanmar, 138,366 deaths
- 2005, Kashmir earthquake, Pakistan/India, 100,000 deaths
- 2008, Sichuan earthquake, China, 87,587
- 2003, heat wave, across Europe, > 70,000 deaths
- 2003, Bam earthquake, Iran, 26,271 deaths
- 2011, earthquake and tsunami, Japan, 15,889 deaths
- 2015, earthquake in Nepal, >9,000 deaths
- 2013, typhoon Haiyan, Philippines, China, Vietnam, 6,340 deaths

However, over a longer period, a different picture emerges. The following graphs (Figures 1.6–1.9), based on recent data obtained from the EM-DAT database, show a dramatic increase in the number of disasters over the last century. Some of the initial messages are:

- The amount of disasters occurring each year is on the increase.
- The number of people killed by disasters has thankfully decreased, but the number of people being affected is generally on the increase.
- The number of people killed and affected by 'technological' disasters has increased.

Figure 1.6 Total number of disasters associated with natural hazards 1915–2015 (*Source:* The OFDA/CRED EM-DAT International Disaster Database).

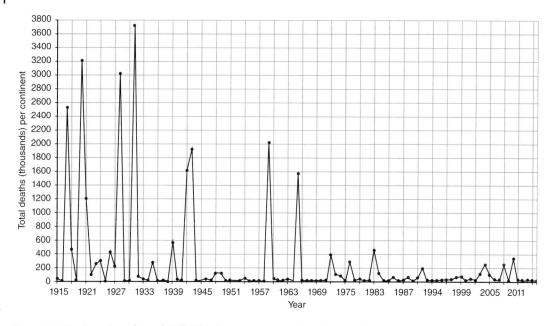

Figure 1.7 Total number of people killed by disasters associated with natural hazards 1915–2015 (*Source:* The OFDA/CRED EM-DAT International Disaster Database).

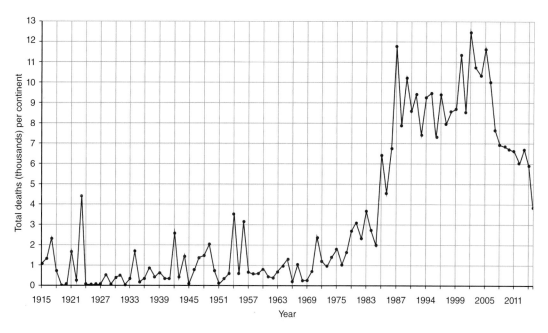

Figure 1.8 Total number of people killed by technological disasters 1915–2015 (*Source:* The OFDA/CRED EM-DAT International Disaster Database).

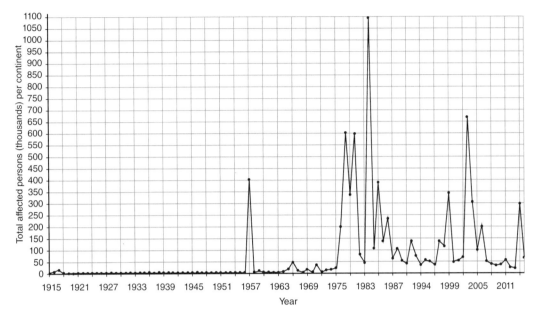

Figure 1.9 Total number of people affected by technological disasters 1915–2015 (*Source:* The OFDA/CRED EM-DAT International Disaster Database).

The trends in the increasing numbers of disasters and their impacts in terms of people being adversely impacted are due to a variety of reasons. In some cases, the number of disasters is actually increasing – for example, technological disasters such as traffic accidents and industrial incidents will increase as the numbers of vehicles and factories increase. These observations are most likely associated to countries where health and safety regulations are not very stringent or their implementation is not being suitably monitored or enforced.

Disasters associated with natural hazards could have been assumed to be constant, but as discussed earlier in this chapter, it is now generally accepted that the world's climate is changing and thus influencing the amount of disasters related to hydro-meteorological hazards such as floods and storms (IPCC, 2014). However, the increase in the number of people affected had already started to grow even before the more obvious signs of climate change, and this factor may have been exacerbated by the relatively rapid rise in the global (and increasingly urbanised) population over the last century. Some hazards, such as earthquakes, are not subject to the immediate impacts of climate change and thus the annual frequency of seismic activity has not significantly changed over the last 50 years (see Figure 1.10). Some of the increases may be due to better reporting and international awareness of disasters. The 24-hour news media and changes in the way nature and death are viewed have helped to make people more aware of how many disastrous events and associated emergencies are occurring globally.

1.6 Economic Losses

One clear increase over the last 50 years has been the rise in the cost of disasters (Figure 1.11). However, this has to be coupled with the fact that economic development has resulted in greater embodied value in buildings and infrastructure, so the re-building costs will increase in line with this. This does not necessarily mean that the incidence or the severity of the disasters has got greater, but the impact on the economy has. Low-income countries appear to suffer less in terms of economic

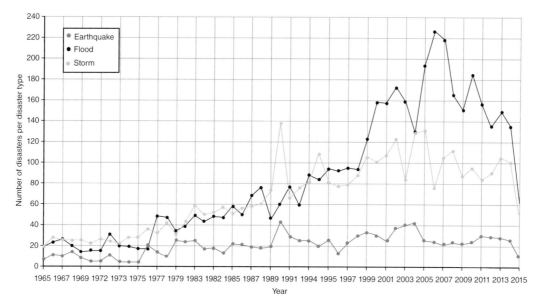

Figure 1.10 Total number of disasters associated with different types of natural hazards 1965–2015 (*Source:* The OFDA/CRED EM-DAT International Disaster Database).

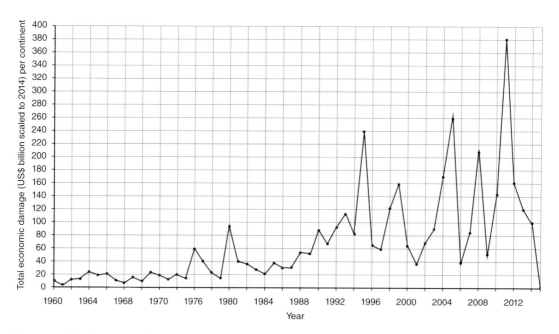

Figure 1.11 Total economic damages caused by disasters associated with natural hazards between 1960–2015 (values normalized to 2014 US$) (*Source:* The OFDA/CRED EM-DAT International Disaster Database).

loss, but that is because they may have had less (extensive/sophisticated) infrastructure to start off with. Therefore, understanding what proportion of economic loss has occurred would be better than merely comparing absolute monetary values.

The UN's Global Assessment Report (UNISR, 2015a) suggests using a new metric – 'life years' – that describes the time required to produce economic development and social progress. The loss of human life years, be it through disasters, disease or accidents, is therefore a way of measuring setbacks to social and economic development (Figure 1.12).

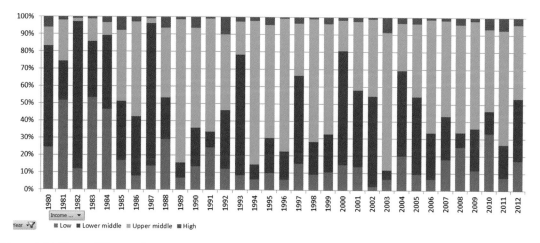

Figure 1.12 Share of life years lost across income groups (*Source:* UN, 2015).

Thinking Point 1.2

Do Not Underestimate 'Mother Nature'

Too often we only focus on the primary effects of the natural hazards, that is, those that occur as a result of a hazard itself. For example, water damage during a flood or collapse of buildings during an earthquake. However, we should not forget that most natural hazards can have larger impacts, which can last for prolonged periods of time. Secondary Effects occur only because a primary effect has caused them; for example, fires ignited as a result of earthquakes, disruption of electrical power and water services as a result of a hurricane, or flooding caused by a landslide into a lake or river. Tertiary Effects are long-term effects that are set off as a result of a primary event. These include loss of habitat caused by a flood, permanent changes in the position of river channel caused by flood, crop failure caused by a volcanic eruption etc.

In addition, a chain reaction, that is, the occurrence of a natural hazard triggered by another hazard, can also take place. Thus, on 26 December 2004, an earthquake ruptured an 800-mile length of the sea floor from northern Sumatra to the Andaman Islands. A massive series of waves rolled across the Indian Ocean. Together, the earthquake and tsunami took more than 200,000 lives in 11 countries. On the morning of 29 November 1975, a magnitude-7.2 earthquake struck the Big Island of Hawaii. Less than 45 minutes later, Kilauea Volcano starting erupting and lasted for 17 hours. The small volume of magma and brief duration suggest that the eruption was triggered by the earthquake. On 18 May 1980, a magnitude 5.1 earthquake triggered the collapse of the north side of Mount St. Helens, resulting in the largest landslide ever recorded. These examples show that hazards cannot be studied in isolation and that multi-hazard approaches should always be carried out.

1.7   The Potential Roles of the Construction Sector in DRR

During the last few decades documented increases of disastrous events have combined with theoretical developments to necessitate a fresh approach to the way in which disasters are managed. Emphasis has moved away from disaster relief and emergency preparedness, towards a more sustainable approach incorporating hazard mitigation and effective risk management (Bosher & Dainty, 2011). It is, therefore, clear that future construction practise needs to be more sensitive to mitigate the impacts of a wide range of hazards. This needs to be achieved through proactive measures. These proactive measures are likely to have a bearing on the professional training (formal and informal) and day-to-day activities of a vast range of construction practitioners; it is this range of 'activities' that is the central focus of this textbook.

It has been argued that designing and constructing a resilient built environment demands an in-depth understanding of the expertise and knowledge of how to avoid and mitigate the effects of threats and hazards (Hamelin & Hauke, 2005). The most influential disciplines that are expected to most significantly take on board DRR activities as part of their everyday duties are the design, engineering and construction professions (these will be detailed in Chapter 8).

Among other requirements, the United Nations' Hyogo Framework for Action 2005–2015 (UNISDR, 2005) and the more recent Sendai Framework for Action 2015-2030 (UNISDR, 2015b) have called on governments to mainstream disaster risk reduction considerations into planning procedures for major construction projects. Fundamentally, there is now more recognition that we (collectively) need to change the way we develop our cities, infrastructure and buildings. It is being argued that this cannot be achieved by merely mainstreaming DRR into construction practice, but as Prof Allan Lavell suggested at the 2015 Sendai conference, we need to make DRR part of the 'developmental DNA'.

What emerges is a picture of a built environment under increasing threat from a multiplicity of different hazards and threats, some well-established but difficult to mitigate, others more emergent and hence somewhat unpredictable (Bosher & Dainty, 2011). Arguably, these hazards are likely to become more significant in future years, and so it has become incumbent upon those responsible for planning, designing and constructing the built environment today to take account of these threats as a core part of their professional activity. It is the decisions taken now that will determine the burden that future generations inherit with regards to their resilience to a range of hazards. Therefore, the efficient planning, designing and constructing of resilience now will lessen the need for expensive retrofitted measures in the future.

1.8   Scope of the Book

This book examines multi-hazard and threat adaptation issues from a construction sector perspective, whilst also considering the relevant non-structural elements of multi-hazard/threat adaptation. The content encompasses international contexts, natural hazards (i.e., hydro-meteorological and geological) and some human-induced threats (i.e., unregulated urban planning and poor quality construction). The book aims to provide a multi-disciplinary perspective, by providing a multi-faceted introduction to how a broad range of risk reduction options can be mainstreamed into the construction decision-making processes and ultimately so that DRR can indeed become part of the 'developmental DNA'.

This book is an introduction and will intentionally cover broad aspects of DRR and not detailed specifics; that is, it will not attempt to supplant books that cover specific and detailed approaches such as 'earthquake engineering', 'flood risk management' and 'hazard mapping using GIS' but will signpost the reader to these types of publications if more technical details are required. The book presents the reader with a suitably broad range of options, and thus it is not intended to be a handbook of prescriptive 'solutions'; it is hoped that the contents of the book will inspire a broad range of construction practitioners to incorporate DRR thinking and innovations into their everyday practice.

1.9 Structure of the Book

Initially the reader will be introduced to key disaster risk concepts and terms that are applicable across disciplines and the global context. Following this, the core principles of effective risk management are explained; these consist of five elements that should be performed more or less, in the following order (Bosher, 2014, after BSI, 2009):

1) Identify, characterise, and assess natural hazards and human-induced threats
2) Assess the vulnerability of critical assets to specific hazards and threats
3) Determine the risk (i.e., the expected consequences of specific hazards/threats on specific assets)
4) Identify ways to reduce those risks
5) Prioritise risk reduction measures

These risk management principles will then be explained and illustrated with examples in the context of a broad range of the most prominent natural hazards and human-induce threats as outlined in the following sections:

- Hydro-meteorological hazards (floods, hurricanes, tornadoes)
- Geological hazards (earthquakes, landslides and tsunamis)

The following chapters are punctuated with strategically placed '**Thinking Points**' (designed to allow reflection as well as to generate discussions in lectures) and '**Case Studies**' (to help illustrate specific examples of disasters, their causes and/or risk reduction options). The book will conclude with an explanation of some key principles that can be adopted and adapted and a discussion of why DRR is an important, if not undervalued, component of sustainable development. The final chapters will provide some reflection on the key research and practical challenges.

References and Suggested Reading

Alexander D. E., (2013), 'Resilience and disaster risk reduction: An etymological journey', *Natural Hazards and Earth System Sciences*, Vol. 1: 1257–1284

Alexander D.E., (2002) *Principles of Emergency Planning and Management*, Harpenden: Terra Publishing and New York: Oxford University Press.

Bosher L.S., (2014), 'Built-in resilience' through Disaster Risk Reduction: Operational issues', *Building Research & Information*, Vol. 42, No. 2, pp. 240–254

Bosher L.S., (ed.), (2008) *Hazards and the Built Environment: Attaining Built-in Resilience*, Taylor and Francis, London

Bosher L.S. and Dainty A.R.J., (2011) 'Disaster risk reduction and 'built-in' resilience: Towards overarching principles for construction practice', *Disasters: The Journal of Disaster Studies, Policy and Management*, Vol. 35, No 1, pp. 1–18

Bosher L.S., Carrillo P.M., Dainty A.R.J., Glass J., and Price A.D.F., (2007) 'Realising a resilient and sustainable built environment: Towards a strategic agenda for the United Kingdom', *Disasters: The Journal of Disaster Studies, Policy & Management*, Vol. 31, No. 3, pp. 236–255

British Standards Institution, (2009). *Risk management: Principles and guidelines.* London: British Standards Institution Group

Burton I., Kates R.W., and White G., (1993),*The Environment as Hazard: Second Edition*, Guilford Press, London

Church, J.A., P.U. Clark, A. Cazenave, J.M. Gregory, S. Jevrejeva, A. Levermann, M.A. Merrifield, G.A. Milne, R.S. Nerem, P.D. Nunn, A.J. Payne, W.T. Pfeffer, D. Stammer and A.S. Unnikrishnan, (2013): 'Sea Level Change'. In: *Climate Change 2013: The Physical Science Basis. Contribution of Working Group I to the Fifth Assessment Report of the Intergovernmental Panel on Climate Change* [Stocker, T.F., D. Qin, G.-K. Plattner, M. Tignor, S.K. Allen, J. Boschung, A. Nauels, Y. Xia, V. Bex and P.M. Midgley (eds.)]. Cambridge University Press, Cambridge, United Kingdom and New York, NY, USA, pp. 1137–1216

EM-DAT (2016) *The OFDA/CRED International Disaster Database*, Université Catholique de Louvain, Brussels, http://www.emdat.be/natural-disasters-trends

Gaillard, J.C. (2010), 'Vulnerability, capacity and resilience: Perspectives for climate and development policy', *Journal of International Development*, 22: 218–232

Guha-Sapir D, Hoyois P. and Below. R. (2013), *Annual Disaster Statistical Review 2012:The Numbers and Trends.* CRED, Université Catholique de Louvain, Brussels

Hamelin, J-P. and Hauke B., (2005), *Focus Areas: Quality of Life—Towards a Sustainable Built Environment*, European Construction Technology Platform, Paris.

IPCC, (2014), 'Summary for policymakers'. In: *Climate Change 2014: Impacts, Adaptation, and Vulnerability. Part A: Global and Sectoral Aspects. Contribution of Working Group II to the Fifth Assessment Report of the Intergovernmental Panel on Climate Change* [Field, C.B., V.R. Barros, D.J. Dokken, K.J. Mach, M.D. Mastrandrea, T.E. Bilir, M. Chatterjee, K.L. Ebi, Y.O. Estrada, R.C. Genova, B. Girma, E.S. Kissel, A.N. Levy, S. MacCracken, P.R. Mastrandrea, and L.L. White (eds.)]. Cambridge University Press, Cambridge, United Kingdom and New York, NY, USA, pp. 1–32

Kelman I., Mercer J. and Gaillard J.C., (Eds.), (2017), *Routledge Handbook of Disaster Risk Reduction and Climate Change Adaptation*, Routledge, London

Lewis J., (1999), *Development in Disaster-prone Places: Studies of Vulnerability*, Intermediate Technology Publications, London

O'Keefe, P., K. Westgate and B. Wisner (1976), 'Taking the naturalness out of natural disasters', *Nature*, 260. pp. 566–567

Oliver-Smith A. and Hoffman S.M., (2002), '*Why Anthropologists Should Study Disasters'*, in *Catastrophe and Culture: The Anthropology of Disaster*, School of American Research Press, Santa Fe

Pelling M., (2003), *The Vulnerability of Cities: Natural Disasters and Social Resilience*, Earthscan, London

Penuel K. B. and Statler M. (2011), *Encyclopedia of Disaster Relief* (Vol. 2), SAGE, Los Angeles

Smith K., (2013), *Environmental Hazards: Assessing Risk and Reducing Disaster*, 6th Edition, London, Routledge

Uitto, J.I. and Shaw, R. eds., (2015), *Sustainable Development and Disaster Risk Reduction*, Springer

UNISDR (2015a) *Global assessment report on disaster risk reduction 2015*, available at: http://www.preventionweb.net/english/hyogo/gar/2015/en/gar-pdf/GAR2015_EN.pdf

UNISDR (2015b), *Sendai Framework for Disaster Risk Reduction 2015-2030*, United Nations International Strategy for Disaster Reduction, Geneva http://www.unisdr.org/files/43291_sendaiframeworkfordrren.pdf

UNISDR (2005), *Hyogo Framework for Action 2005-2015: Building the Resilience of Nations and Communities to Disasters*, United Nations International Strategy for Disaster Reduction, Geneva http://www.unisdr.org/eng/hfa/docs/Hyogo-framework-for-action-english.pdf

White G.F., (1945), *Human adjustment to floods, Research Paper 29*, Department of Geography, University of Chicago, 225 pp

Wisner B., Blaikie P., Cannon T., and Davis I. (2004), *At Risk: Natural Hazards, People's Vulnerability, and Disasters: Second Edition*, London, Routledge

Wisner, B., Gaillard, J.C. and Kelman, I. eds., (2012), *Handbook of hazards and disaster risk reduction and management*, Routledge, London

2

Disaster Risk Reduction

The more governments, UN agencies, organizations, businesses and civil society understand risk and vulnerability, the better equipped they will be to mitigate disasters when they strike and save more lives.

— Ban Ki-moon, United Nations Secretary-General

2.1   Learning Objectives

Disasters can often appear to be caused by natural hazards but typically there are key underlying socio-economic and development factors that effectively turn the hazards into a disaster. Thus the likelihood, magnitude and impact of a disastrous event often depends on the decisions made when planning, constructing and maintaining the built environment (CIB 2016). These choices relate to the ways natural systems are treated, the location and methods used in construction practice as well as the systems of governance, finance and education. Each decision and action can make the built environment – and its inhabitants – more vulnerable or more resilient to disasters.

By the end of this chapter you will have learnt about the:

- Key disaster risk reduction concepts and terms;
- Main international approaches to Disaster Risk Reduction (DRR);
- Phases of disaster and risk management;
- Range of DRR options

2.2   Key DRR Concepts and Terms

Disasters can be reduced by decreasing the exposure to hazards, lessening vulnerability of people and property, the sensible management of land and the environment, and by improving preparedness and early warning for adverse events. Disaster risk reduction includes disciplines like disaster management, hazard mitigation and emergency preparedness, but DRR is also considered an integral part of sustainable development (see more in Chapter 9).

> **Disaster risk reduction** is the concept and practice of reducing disaster risks through systematic efforts to analyse and reduce the impacts of disasters.

Disaster Risk Reduction for the Built Environment, First Edition. Lee Bosher and Ksenia Chmutina.
© 2017 John Wiley & Sons Ltd. Published 2017 by John Wiley & Sons Ltd.

Figure 2.1 Devastation caused by the 2010 Haiti earthquake (*Source:* US Embassy Kathmandu on Flickr).

In order to fully appreciate the concept of Disaster Risk Reduction, the following terms and concepts should first be understood:[1]

> **Risk** is a probability of an event and its negative consequences.

The word '**risk**' has two distinctive connotations: in general usage the emphasis is usually placed on the concept of chance or possibility (for instance, 'the risk of an accident'); whereas in technical settings the emphasis is usually placed on the consequences, in terms of 'potential losses' for some particular cause, place and period. People do not necessarily share the same perceptions of the significance and underlying causes of different risks.

Disaster risk can be defined as the potential disaster losses in lives, health status, livelihoods, assets and services, which could occur to a particular location, community or society over a specified time period. This definition reflects the concept of disasters as the outcome of continuously present conditions of risk. Disaster risk comprises different types of potential losses which are often difficult to quantify. Nevertheless, knowing the prevailing hazards and the patterns of population and socio-economic development, disaster risks can be assessed, mapped, and therefore reduced in broad terms at least (as will be discussed later in this chapter).

Some risks are considered *acceptable*: these are potential losses that a society or community considers tolerable in the context of existing social, economic, political, cultural, technical and

---

1 Here we use the definitions provided by the UNISDR that are commonly adopted in DRR activities.

environmental conditions. There are also risks that can be impossible to manage – these are called *'residual risks'*.

Intensive risk is a characteristic of cities (i.e., densely urbanised and populated areas); it is associated with the exposure of large concentrations of people and economic activities to intense hazard events, which can lead to potentially catastrophic impacts involving high mortality and asset loss. *Extensive risk* is associated with the exposure of dispersed populations (mainly rural or semi-urban) to repeated or persistent hazard conditions of low or moderate intensity, often of a highly localised nature, which can lead to debilitating cumulative disaster impacts.

> **Vulnerability** is the characteristics and circumstances of a community, system or asset that make it susceptible to the damaging effects of a hazard.

Vulnerability has many aspects including various physical, social, economic, and environmental factors. Examples may include poor design and construction of buildings, inadequate protection of assets, lack of public information and awareness, limited official recognition of risks and preparedness measures, and disregard for wise environmental management. It can vary significantly within a community and over time, and is specific to each location even if other factors (e.g., economic development) are similar. Vulnerability is most often associated with poverty, but it can also arise when people are isolated, insecure and defenceless in the face of risk, shock or stress. People can differ in their exposure to risk as a result of their social group, gender, ethnicity, religion, age and health, and so on.

> **Mitigation** is the lessening or limitation of the adverse impacts of hazards and related disasters.

Mitigation implies that whilst adverse impacts of hazards often cannot be prevented fully, their severity can be substantially reduced by various strategies and actions. Mitigation measures encompass structural (e.g., engineering techniques, hazard-resistant construction) and non-structural approaches (e.g., improved environmental policies, public awareness). It should be noted that in the context of climate change policy, 'mitigation' is defined differently: the term is used for the reduction of greenhouse gas emissions that are the source of climate change.

> **Preparedness** is the knowledge and capacities developed by governments, professional response and recovery organisations, communities and individuals to effectively anticipate, respond to, and recover from, the impacts of likely, imminent or current hazard events or conditions.

Preparedness takes place within the context of disaster risk management and aims to build the capacities needed to efficiently manage all types of emergencies and achieve orderly transitions from response through to sustained recovery. Preparedness is based on a comprehensive analysis of disaster risks and establishment of effective early warning systems. It includes such activities as contingency planning, stockpiling of equipment and supplies, the development of arrangements for coordination, evacuation and public information, associated training and field exercises, as well as household measures for evacuation (see Figure 2.2 for a commercially available evacuation bag). In order to be effective, preparedness actions must be supported by formal institutional, legal and budgetary capacities and implemented thoroughly at the national, regional and local levels.

Figure 2.2 Example of a two-person 72-hour emergency kit go bag (*Source:* Reproduced with permission of EVAQ2).

> **Recovery** is the restoration and improvement, where appropriate, of facilities, livelihoods and living conditions of disaster-affected communities, including efforts to reduce disaster risk factors.

Recovery includes rehabilitation and reconstruction (Figure 2.3) that should begin soon after the emergency phase has ended, and should be based on pre-existing strategies and policies that facilitate clear institutional responsibilities for recovery action and enable public participation. Recovery programmes, coupled with the heightened public awareness and engagement after a disaster, offer a valuable opportunity to develop and implement disaster risk reduction measures and to apply the 'build back better' principle (this will be discussed in more detail in Chapter 8).

> **Resilience** is the ability of a system, community or society exposed to hazards to resist, absorb, accommodate to, and recover from the effects of a hazard in a timely and efficient manner, including through the preservation and restoration of its essential basic structures and functions.

Resilience is a rather complex and widely debated concept that is increasingly being used in DRR (see Thinking Point 2.1). Resilience is often used to describe the degree to which the country/business/community has the necessary resources and is capable of organising itself both prior to, and during, times of need. The term 'resilience' is increasingly adopted by both policy makers and practitioners in the field of DRR; however, both the conceptual clarity and practical relevance of this term

Figure 2.3 Reconstruction in Nepal after the 2015 earthquakes (*Source:* Reproduced with permission of Rohit Jigyasu).

are unclear. The original meaning was largely constructed in the field of ecology, and was understood as a measure of the ability of ecological systems to persist in the face of disturbance and maintain relationships between different elements of the system; this idea has been recently adapted (and, while doing so, diluted and stretched) by many other disciplines, creating ambiguity and uncertainty. Unsurprisingly, this sometimes siloed approach has led to major difficulties in operationalising and applying resilience in the search for more harmonious relationships between the environmental, the social, the institutional and the physical components of resilience (Figure 2.4).

Social resilience includes the ways in which individuals and groups of people deal with crises and develop coping mechanisms and anticipatory adaptation strategies.

Institutional resilience is about the ways that governmental and non-governmental institutions prepare for and manage crises and develop proactive adaptation/mitigation strategies.

Physical resilience of buildings and critical infrastructure refers to how the public and private sectors (including utilities and transport companies) manage crises and develop proactive adaptation/mitigation strategies so that disastrous events can be avoided or at least the impacts and disruption minimised.

Environmental resilience is about society, institutions and the commercial sectors supporting the capacity of the natural environment to cope with environmental changes.

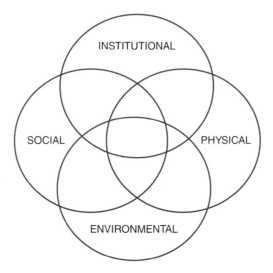

Figure 2.4 The components of resilience.

Despite the lack of clarity about what 'resilience' actually means, a growing number of governmental and non-governmental reports are mentioning resilience. Similarly, the term has become common among local authorities, construction stakeholders and emergency services.

2.3   International Approaches to DRR

DRR sets out to bridge the gap between development and livelihood security, as development can only be sustained if there is a clear understanding of and response to the negative impact of disasters. DRR interventions seek to assist in the development of this understanding, to support livelihoods and to protect assets. As an increasing number of people are being affected by natural hazards, there has been a growing recognition by governments and organisations that building resilience and reducing disaster risk should be central to their everyday activities.

2.3.1   Milestones in History of DRR

Over the decades, DRR as a field has moved from a narrowly perceived technical discipline to a broad-based global movement focused on sustainable development.

The first steps towards mainstreaming DRR were taken in 1960s when the UN General Assembly (GA) adopted a number of minor measures regarding severe 'natural disasters': Member States were requested to inform the Secretary-General of the type of emergency assistance they are in a position to offer (See Case Study 2.1 for more details on the role of the UN in DRR).

During 1970s and 1980s, a number of important DRR events and agreements that helped improving assistance in cases of 'natural disasters' took place, including:

1970: The General Secretary was invited to submit recommendations on *pre-disaster planning* at the national and international levels, and the application of technology to, and scientific research for, the prevention and control of natural disasters, or a mitigation of the effects of such disasters.

Thinking Point 2.1

What Does 'Resilience' Actually Mean?

The term *'resilience'* has recently been widely used, but the word itself isn't new and has a very long history. Current debates on resilience mainly refer to the seminal work of Holling and his landmark paper on ecosystems published in 1973, and it is often believed that resilience has been developed as an ecological concept and applied to human-environment systems; some of this work actually draws upon the even earlier work of scientists such as Errington (1953) and Blum (1968). In reality, the word has a much longer history: its root 'resilire'/ 'resilio' comes from Latin for 'bounce' (hence the idea of 'bouncing back'). Seneca uses the term in a sense of 'to leap'. Much later, the term passed into Middle French (*résiler*), in which it came to mean 'to retract' or 'to cancel', and then into English as the verb *resile*, used again in the sense of 'retract', 'return to a former position', or 'desist'. David Alexander (2013) suggests that the first scientific use of the word 'resilience' was by Bacon in 1625 when writing about natural history. Importantly, from 1839, the term was also used to signify the ability to recover from adversity: the word 'resiliency' was used in the sense of the ability to withstand the effects of earthquake with respect to observations made by Americans during the recovery of the city of Shimoda after an earthquake. The first use of the term *resilience* in mechanics appeared in 1858, when the eminent Scottish engineer William J. M. Rankine employed it to describe the strength and ductility of steel beams. At about the same time, further applications of the term were being made in coronary surgery, anatomy and watchmaking. In the 1950s, the term resilience started to be used in psychology (particularly in relation to the psychiatric problems of children) and it finally became popular in field of disaster management in the late 1980s. Since then resilience has become an idea used on many different scales with many different intentions and with a very wide extension. It includes a range of components, from international aid and leadership to resistance and security to sustainability and community well-being. It is used to connect discourses of separate professionals but in equal measure it may confuse them by conflating many meanings. This makes it impossible to decide whether a specific state is resilient or not, and to find out how a resilient state can be achieved. The principle of resilience, therefore, should be adopted with prudence, and the long-term consequences of interrelated variables must be considered. For more information about this topic it is worth reading the following paper by David Alexander:

Alexander D. E. (2013), 'Resilience and disaster risk reduction: An etymological journey'. *Natural Hazards and Earth System Sciences*, 1, 1257–1284

1971: *Creation of the United Nations Disaster Relief Office (UNDRO)*, which was charged with promoting the study, prevention, control and prediction of 'natural disasters'; assisting in providing advice to governments on pre-disaster planning; and helping to improve national disaster warning systems.

1972: The importance of *preventive measures, disaster contingency planning and preparedness* is reaffirmed by the GA.

1979: Creation of *Office of the United Nations Disaster Relief Coordinator* (through the UNDP) responsible for the development of the New International Development Strategy to take into account matters concerning disaster relief, preparedness and prevention.

The 1990s were announced as *International Decade for Disaster Risk Reduction*. It was launched by the United Nations, following the adoption of Resolution 44/236 (22 December 1989). The decade was intended to reduce, through concerted international action, especially in developing countries, loss of life, property damage and social and economic disruption caused by natural hazards.

1990: The GA urges the international community to implement fully *the International Framework of Action of the International Decade for Natural Disaster Reduction (IDNDR)*, to establish national committees and reaffirm the need for the secretariat of the Decade to work in close co-operation with UNDRO.

1994: *The World Conference on Disaster Reduction*, Yokohama, as a result of which the Yokohama Strategy and its Plan of Action was adopted. It provides guidelines for prevention, preparedness and mitigation of natural hazards, describes the principles on which a disaster reduction strategy should be based, introduces a plan of action agreed upon by all member states of the United Nations, and gives some guidelines concerning the follow-up of action. It also drives the implementation of early warning systems.

1997: Recognition of the *El Niño* phenomenon (Figure 2.5).

1999: The *IDNDR Programme Forum* is launched recognising need to generate 'a global culture of prevention'.

El Niño is a complex interaction of the tropical Pacific Ocean and the global atmosphere that results in irregularly occurring episodes of changed ocean and weather patterns in many parts of the world, often with significant impacts over many months, such as altered marine habitats, rainfall changes, floods, droughts, and changes in storm patterns.

La Niña is a cooling of the water in the equatorial Pacific, which occurs at irregular intervals, and is associated with widespread changes in weather patterns complementary to those of El Niño, but less extensive and damaging in their effects

---

The **2000s** were also important in terms of developing the further understanding of disasters, vulnerabilities and the establishment of the UNISDR.

2000: The *International Strategy for Disaster Reduction* (ISDR) was established creating an inter-agency task force and inter-agency-secretariat for disaster reduction. Activities included the observance *of the International Day for Disaster Reduction* on the second Wednesday of October.

2002: *The Johannesburg Plan of Action*, implemented at the World Summit on Sustainable Development (WSSD), provided the ISDR with a concrete set of objectives within the sustainable development agenda to which both the Inter-Agency Task Force on Disaster Reduction and the UN/ISDR secretariat, along with partners, would increasingly turn their attention and capacities to integrating and mainstreaming risk reduction into development policies and processes.

2005: Implementation of the Hyogo Declaration and the *Hyogo Framework for Action* 2005–2015 (HFA). It was the first plan to explain, describe and detail the work that was required from all different sectors and actors to reduce disaster losses. It was developed and agreed on with the governments, international agencies, disaster experts and many others stakeholders bringing them into a common system of coordination. The HFA outlined five priorities for action, and offered guiding principles and practical means for achieving disaster resilience. Its goal was to substantially reduce disaster losses by 2015 by building the resilience of nations and communities to disasters. Generally, it was felt that the HFA was a big move in the right direction but that improvements needed to be made; for instance, it was noted that handling what was primarily a developmental issue with largely relief and humanitarian mechanisms and instruments needed to be reconsidered to ensure that disaster risk reduction plays the role that it must in enabling and safeguarding development gains.

**Typical January - March weather anomalies and atmospheric circulation
during moderate to strong El Niño and La Niña**

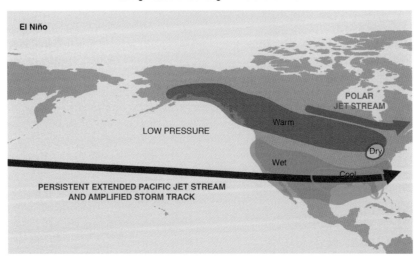

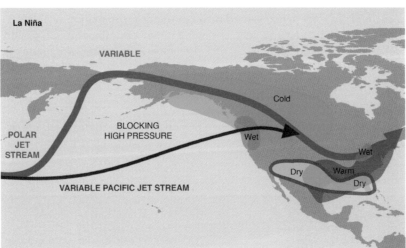

Figure 2.5   El Niño and La Nina conditions (*Source:* WEDC).

2006: *The Global Platform on Disaster Reduction* is established as the successor mechanism of the
Inter-Agency Task Force for Disaster Reduction.

2.3.2   Sendai Framework for Disaster Risk Reduction

The Hyogo Framework for Action (HFA) 2005–2015 was replaced by the Sendai Framework for
Disaster Risk Reduction 2015–2030, that was introduced and negotiated during the third United
Nations World Conference on DRR in March 2015 in Japan (Figure 2.6). The new Framework is a
more nuanced version of the previous HFA, with some overall guiding principles, a set of common
standards, targets, and a legally based instrument for disaster risk reduction.

Case Study 2.1

The Role of UN in DRR

The United Nations Office for Disaster Risk Reduction (UNISDR) was established in 1999. It serves as the focal point in the UN system for the coordination of disaster risk reduction and ensures the implementation of the International Strategy for Disaster Reduction (i.e., the promotion of 'a culture of prevention'). UNISDR coordinates international efforts in DRR and guides, monitors and reports on progress of the implementation of the Hyogo Framework for Action and Sendai Framework for Disaster Risk Reduction (see Case Study 2.2); campaigns and advocates creation of global awareness of DRR benefits and empowers people to reduce their vulnerability to hazards; encourages for greater investments in risk reduction actions to protect people's lives and assets including climate change adaptation, education and increased participation in the decision making process; and informs and connects people by providing practical services and tools (see Case Study 2.4).

Search terms: UNISDR, Making Cities Resilient Campaign; Hyogo Framework for Action; Sendai Framework for Disaster Risk Reduction

---

Taking into account the experience gained through the implementation of the HFA, and largely maintaining its expected outcome and goal, the new Framework offers more focus on multi-stake-holder approaches at local, national, regional and global levels in the following four priority areas:

1) Understanding disaster risk;
2) Strengthening disaster risk governance to manage disaster risk;
3) Investing in disaster risk reduction for resilience;
4) Enhancing disaster preparedness for effective response, and to "Build Back Better" in recovery, rehabilitation and reconstruction.

Figure 2.6 Sendai UN Conference on DRR in March 2015 (*Source:* Lee Bosher).

Figure 2.7 Making cities resilient campaign (*Source:* UNISDR).

It points out the role of various stakeholders including not only the state but also civil society (and the most vulnerable population groups), academia, private businesses, and media as well as international organisations. It also emphasises the role of the international cooperation and global partnership. Fundamentally, there is now more recognition that we (collectively) need to change the way we develop our cities, infrastructure and buildings, by not merely mainstreaming DRR into practice but by making DRR part of the 'developmental DNA'.

2.3.3 'Making Cities Resilient' Campaign

The 2010-2015 World Disaster Reduction Campaign "Making Cities Resilient" (Figure 2.7) addresses issues of local governance and urban risk while drawing upon previous UNISDR Campaigns on safer schools and hospitals, as well as on the sustainable urbanisation principles developed in the UN-Habitat World Urban Campaign 2009–2013.

The objectives of the Campaign are to support sustainable urbanisation by promoting resilience activities, increasing local level understanding of disaster risk, and encouraging commitments by local and national governments to make disaster risk reduction and climate change a policy priority, and to bring the global Hyogo Framework for Action (and now the Sendai Framework) closer to local needs. The Campaign Framework was developed through a one-year consultative process (2008–2009), which culminated with a conference in Incheon City in August 2009, opened by United Nations Secretary-General Ban Ki-moon and led by UNISDR and United Cities and Local Governments (UCLG). The outcomes of that meeting were captured in the 'Incheon Declaration', which rooted the campaign's development and strategy. The Campaign is guided by three central themes: know more, invest wiser, and build safer. These have formed the basis for the development and deployment of practical tools and resources to help support and empower local governments to build their resilience to disasters and are based on the five priorities of the Hyogo Framework for Action.

Over 3,400 cities had signed up to this campaign by early 2017; these cities are required to use the Local Government Self-Assessment Tool, which helps to set baselines, identify gaps, plan actions and have comparable data across local governments, within the country and globally, to measure advancements over time (see Case Study 2.2).

Case Study 2.2

Ten-Point Checklist – Essentials for Making Cities Resilient

A ten-point checklist was developed in line with the five priorities of the Hyogo Framework for Action 2005–2015:

1) Put in place **organisation and coordination** to understand and reduce disaster risk, based on participation of citizen groups and civil society. Ensure that all departments understand their role to DRR and preparedness.
2) **Assign a budget** for DRR and provide incentives for homeowners, low-income families, communities, businesses and public sector to invest in reducing the risks.
3) Maintain **up-to-date data** on hazards and vulnerabilities, prepare risk assessments and use these as the basis for urban development plans and decisions. Ensure that this information is publicly available.
4) Invest in and **maintain critical infrastructure** that reduces risk adjusted where needed to cope with climate change.
5) Assess the **safety of all schools and health facilities** and upgrade these as necessary.
6) Apply and enforce realistic, risk compliant **building regulations and land use planning** principles. Identify safe land for low-income citizens and develop upgrading of informal settlements, wherever feasible.
7) Ensure **education programmes and training** on DRR are in place.
8) **Protect ecosystems and natural buffers** to mitigate floods, storm surges and other hazards. Adapt to climate change by building on good risk reduction practices.
9) Install **early warning systems and emergency management** capacities and hold regular public preparedness drills.
10) Ensure that the **needs of the victims** are placed at the centre of reconstruction.

Search terms: UNISDR, Making Cities Resilient Campaign; checklist

---

2.4 Community Resilience

Different communities have different hazards, vulnerabilities and capacities depending on their context (Figure 2.8). Determining actual risk awareness is a complex matter requiring an understanding of local culture, social and spiritual norms and perceptions, history, education and personality. Similarly to the term 'resilience', it is therefore extremely hard to agree on a definition of 'community resilience' (See Thinking Point 2.2).

> **Community resilience** is a measure of the sustained ability of a community to utilize available resources to respond to, withstand, and recover from adverse situations.

Communities are often defined in terms of space (i.e., groups of people living in the same area or close to the same risks), thus overlooking other significant dimensions of 'community' which are to do with common interests, values, activities and structures. In reality, communities are complex: whilst being in the same geographical area, they may differ in wealth, social status and employment activities. Individuals can be members of several communities at the same time, linked to each by different factors such as location, occupation, economic status, gender, religion or recreational interests.

Figure 2.8 Promoting community resilience to extreme weather in Cambodia. (*Source:* USAID Asia).

No community can ever be completely safe from hazards and threats, and different layers of resilience are needed to deal with different kinds and severities of risk, shock, stress or environmental change. It may be helpful to think of a disaster resilient or disaster-resistant community as the safest possible community that can be created in a natural hazard context, minimising its vulnerability by maximising the application of DRR measures. Resilient communities commonly share a number of characteristics:

1) Knowledge and well-being: the community has the ability to assess, manage and monitor its risks and learn new skills as well as build on past experiences
2) Governance and organisation: the community has the capacity to identify problems, establish priorities and act accordingly.
3) Connection with external stakeholders: the community has an established relationship with external actors who provide a wider supportive environment, and supply goods and services when needed.
4) Operation and maintenance of infrastructure and services: the community has strong housing, transport, electrify, water and sanitation systems. It has the ability to maintain, repair and renovate them.
5) Economic well-being: the community has a diverse range of employment opportunities, income and financial services. It is flexible, resourceful and has the capacity to accept uncertainty and respond (proactively) to change by ensuring business continuity.
6) Sustainability: the community recognises the value of natural assets and has the ability to protect, enhance and maintain them.

Thinking Point 2.2

Resilient 'Communities'

Disasters create impacts at a local level, with the local communities being at the forefront of both the immediate effects of a disaster and the initial, emergency response, which, experience has shown, is crucial for saving the most lives. It is therefore critical to enhance community resilience to hazards and threats: this point is emphasised in the Hyogo Framework for Action and Sendai Framework on Disaster Risk Reduction. Resilient communities are also at the core of the Making Cities Resilient Campaign (see Case Study 2.2).

UNISDR states that 'The resilience of a community in respect to potential hazard events is determined by the degree to which the community has the necessary resources and is capable of organising itself both prior to and during times of need'. However the main challenge is that there is no panacea for how to achieve community resilience. Every community has to evaluate its strengths and weaknesses, and assess their vulnerabilities and risks. To further complicate matters, the term community is not always clearly understood. For instance the word 'community' has a number of different meanings which are context specific; 'community' could refer to a geographically bounded notion such as a village or neighbourhood or it may actually refer to (less visible or geographically bounded) alignments based on religious, racial or cultural grounds.

The Community and Regional Resilience Institute (CARRI) states that it is impossible to come up with one definition, which will be useful for all the 'communities', however, the resilient community (however it is defined) should embed the following core concepts:

- Resilience is an inherent and dynamic attribute of the community and should exist throughout the life of the community as it cannot be absolute;
- Adaptability is at the core, and can occur either in response to or in anticipation of a crisis.
- Any adaptation must improve the community, that is, must result in a positive outcome for the community relative to its state after experiencing adversity.
- Resilience should be defined in a manner that enables useful predictions to be made about a community's ability to recover from adversity. This will enable communities to assess their resilience and take action to improve it if necessary.

2.5   Risk Management

Disaster risk management is a systematic process of using administrative directives, organisations, and operational skills and capacities to implement strategies, policies and improved coping capacities in order to lessen the adverse impacts of hazards and the possibility of a disaster. Disaster risk management aims to avoid, lessen or transfer the adverse effects of hazards through activities and measures for prevention, mitigation and preparedness.

2.5.1   Phases of Disaster Risk Management

In many parts of Europe, emergency management since the Second World War and the Cold War has focused primarily on preparedness. Often this involved preparing for enemy attack or some form of civil emergency. Preparedness for emergencies required identifying resources and expertise in advance, and planning how these can be used in a crisis. However, preparedness is only one phase of emergency or risk management. The fuller range of phases have often (rather myopically) been illustrated as a cycle of phases, in which a disaster event invariably leads to another disaster event (see Thinking Point 2.3).

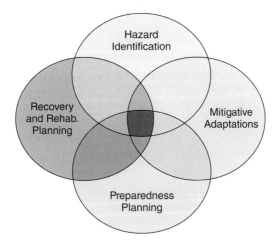

Figure 2.9 Phases of disaster risk management.

Current thinking defines four phases of Disaster Risk Management (DRM) in a more interlinked manner where disaster impacts are not shown as it is felt that a disaster impact should not really be required to instigate thinking and actions for DRM (Figure 2.9):

- *Hazard Identification* - Identification of potentially damaging physical events, phenomenon or human activity.
- *Mitigative Adaptations* ('Hazard Mitigation') - Structural and non-structural measures undertaken to limit the adverse impact of hazards/threats.
- *Preparedness planning* - Activities and measures taken in advance to ensure effective response to disasters

---

Thinking Point 2.3

The 'Disaster Cycle'

Many documents and publications referring to how disasters can be managed mention the 'disaster (management) cycle' as version of which is shown in Figure 2.10. The cycle shown typically indicates a disaster event that is then followed by relief/rehabiliation activies, then reconstruction and hazard mitigation and emergency preparedness but then…... followed by yet another disaster! This implies a cyclical process of actions which always involves a disaster (and efforts to reduce disaster potential and disaster impacts) but results in yet another disaster. This may actually be the case in reality but of course a new way of thinking about the phases of disaster risk management is required.

U.S. comedian Jon Stewart famously picked up on this issue after Hurricane Katrina affected New Orleans in 2005 when he commented on the 'disaster cycle' diagram used by FEMA; stating *"This chart clearly depicting the agencies [sic] responsibilities in the event of a disaster....It begins with a response to a disaster, leads to recovery, mitigation, risk reduction, prevention, preparedness.............. and ends up BACK IN DISASTER! "In truth, FEMA did exactly what they said they were going to do."*

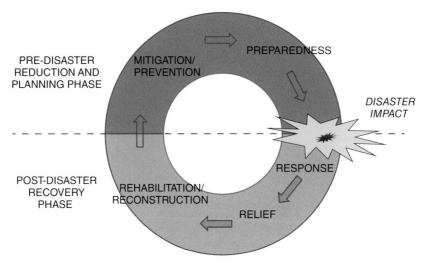

Figure 2.10   Typical illustration of the "disaster cycle."

- *Recovery and rehabilitation* - Decisions and actions taken after a disaster with a view to restoring or improving the pre-disaster living conditions of the stricken community (see Thinking Point 2.4).

Thinking Point 2.4

Phases of a Disaster

It is important to understand the various (often overlapping) phases that normally accompany a disaster, particularly from a point of view of the affected communities (Figure 2.11):

1) *Pre-disaster phase* includes the amount of warning a community receives, which can vary depending on the type of a hazard. Perceived risk varies due to many factors, including previous experience, awareness of the consequences, and so on.
2) *Impact phase* consists of the realisation of the consequences of a hazard. The greater the scope, community destruction, and personal losses associated with the disaster, the greater the psychosocial effects.
3) *Heroic phase* is characterised by high levels of altruism among both survivors and emergency responders. During this phase, search and rescue missions begin, and direct assistance to people and resources arrives. The main goal of this phase is to prevent loss of lives and to minimise property damage.
4) During *the honeymoon phase*, survivors feel a short-lived sense of optimism, as people are appreciative of that help. Survivors hope things will recover quickly and life will return to normal.
5) However as survivors go through an inventory process, they realise the limits of available disaster assistance. Optimism fades, as the realisation of what happens begins to settle in, and disappointment, resentment, anger and frustration become evident. This leads into the *Disillusionment Phase* where survivors are coming to grips with reality of their situation.
6) *Reconstruction phase* takes place after the emergency responders leave and people start adjusting to the new context and take steps to gradually re-establishing the functionality of the space.

Figure 2.11 Phases of disaster (*Source:* Adapted from SAMHSA).

Structural measures: Any physical construction to reduce or avoid possible impacts of hazards, or application of engineering techniques to achieve hazard-resistance and resilience in structures or systems;

Non-structural measures: Any measure not involving physical construction that uses knowledge, practice or agreement to reduce risks and impacts, in particular through policies and laws, public awareness raising, training and education.

2.5.2 Risk Management Elements

Typically risk management consists of five elements, performed more or less, in the following order (adapted from ISO 31000 'Risk management – Principles and guidelines' (British Standards Institution, 2009):

Risk management approach (adapted from ISO 31000)

1) Identify, characterize and assess natural hazards/human induced threats
2) Assess the vulnerability of critical assets to specific hazards/threats
3) Determine the risk (i.e., the expected consequences of specific hazards/threats on specific assets)
4) Identify ways to reduce those risks
5) Prioritize risk reduction measures

---

1) *Identify, characterise, and assess hazards/threats*: The aim of this stage is to begin recognising the threats and hazards to which a location is exposed to. There is a wide range of multi-hazard/ threat identification approaches (many of which can be hazard/ location/ resource specific, and will be discussed in the following chapters) including:
   - Hazard mapping
   - Assessment of recent events

- Assessment of historical events
- Local knowledge
- Scientific research – sensors, computational models

2) *Assess the vulnerability of critical assets to specific hazards/threats*: This is the process of assessing the susceptibility of the intrinsic properties (the structure, materials, construction and planning) to the hazards/threats that can lead to an event with a consequence.

3) *Determine the risk* (i.e., the expected consequences of specific hazards/threats on specific assets): This stage fundamentally aims to answer three questions:

   a) What can happen? ('What can go wrong?')
   b) How likely is it?
   c) What are the consequences? ('How bad could it be?')

   In addition to the above standard three questions, it has been suggested that a fourth question should be added to the list:

   d) How much uncertainty is present in the analysis? (In other words, 'How reliable are the answers to questions 3a–3c?')

Quantitative risk analysis uses probabilistic estimates for all undesired events and risk is determined as the mathematical expectation of the consequences of the undesired events. The aim of a probabilistic logic (or probability logic) is to combine the capacity of probability theory to handle uncertainty with the capacity of deductive logic to exploit structure. The result is a richer and more expressive formalism with a broad range of possible application areas.

Qualitative risk analysis focuses on the likelihood of a hazard(s) occurring and the consequences if a hazard occurs, and using a simple 'Low (L)', 'Moderate (M)', 'High (H)' or 'Extreme (E)' classification, can provide a very visual and easy to understand approximation of the risk scale (Figure 2.12).

4) *Identify ways to reduce those risks*: The aim of this stage is to identify a course of action to address and treat the hazards/ threats and risks associated with them. The hierarchy of risk reduction options (starting with the most preferred in ideal circumstance) is the following:

- **Inherent safety** – eliminate the possibility of hazards occurring (see Thinking Point 2.5)
- **Prevention** – reduce the likelihood of hazards
- **Detection** – measures for early warning of hazards
- **Control** – limiting the size of the hazards
- **Mitigation and adaptation** – protection from the effects of hazards
- **Emergency response** – planning for evacuation and access for emergency services

Likelihood	Consequences				
	Insignificant	Minor	Moderate	Major	Severe
Almost certain	M	H	H	E	E
Likely	M	M	H	H	E
Possible	L	M	M	H	E
Unlikely	L	M	M	M	H
Rare	L	L	M	M	H

Figure 2.12 A typical risk matrix.

Thinking Point 2.5

Is Inherent Safety Ever a Viable Option?

This risk management option is about eliminating the possibility of hazards occurring. But in reality how viable is this? A quick look at Table 2.1 suggests that for most hazards this is not really an option. However there are some circumstances when 'Inherent safety' is possible; can you think of any examples when this is a viable risk management option?

Table 2.1 gives a simplified overview of the risk reduction options that can (or cannot) be applied to various hazards. It is, however, important to bear in mind that the best options will invariably be context specific and consider the possible interactions of differing hazards.

5) *Prioritise risk reduction measures*: Once the potential course of action has been identified, it is important to prioritise the most suitable options. Thus, the objective of this stage is to assist in identifying the most appropriate intervention(s) for a given project. The prioritisation will depend on a number of factors individual to each project; these include (but are not limited to):

- Costs versus benefits of identified interventions
- Technical feasibility
- Social feasibility and acceptability
- Proportionality of identified interventions
- Complementarity with measures introduced to mitigate other hazards
- Corporate social responsibility
- Business continuity
- Legal and statutory requirements

Table 2.1 Overview of the Risk Reduction Measures.

Risk reduction option/Hazard	Natural hazards							Threats	
	Earthquake	Tsunami	Volcano	Landslide	Flood	Storm - cyclone	Storm - tornado	Terrorism	Fire
Inherent safety	N	N	N	#	#	N	N	#	#
Prevention of hazard	N	N	N	#	#	N	N	#	#
Detection of hazard	N	Y	#	Y	#	Y	N	#	Y
Control of hazard	N	N	N	#	#	#	N	#	Y
Mitigation of hazard	Y	#	#	Y	Y	Y	Y	#	Y
Emergency response	Y	#	#	#	Y	Y	#	#	Y

'Y' – there are a range of useful risk options available
'#' – some risk reduction options can be used but they are likely to be of only limited effectiveness
'N' – other than relocating the built asset/community there is little that can be done to reduce this hazard/threat

2.5.3 Existing Guidance

There is an increasingly complex amount of information and guidance on how natural hazards and human-induced threats can be eliminated/reduced/mitigated/designed-out through urban planning and design interventions (see Case Study 2.3). The range of guidance, tools and approaches typically vary depending on the types of hazards/threats that need to be addressed and many are context/country specific in their requirements.

Central to risk management is the international risk management standard *ISO 31000 'Risk management – Principles and guidelines'*, which presents four stages, those being risk identification, assessment, evaluation, and treatment. This framework provides suitable relevance (in functionality and terminology used) globally; it can be applicable to a range of urban contexts (i.e., city, building, organisation); a broad range of practitioners (i.e., planners, architects, security consultants, engineers, local authorities); and covering a range of countries. In addition, this framework allows for the assessment of multi-hazards, thus emphasising that whilst it is not always possible to fully eliminate the risk of all the hazards, it is important to bear in mind that mitigation of one type of risk can enhance another type of risk, if not considered simultaneously.

In the UK, the *Civil Contingencies Act* (see Case Study 8.4) establishes a coherent framework for emergency planning and response ranging from local to national level and defines and roles and responsibilities of different stakeholders during preparedness, response and recovery (but not necessarily on prevention or hazard mitigation).

Some of the guidance is more specific to the construction sector. A key objective of any *building code* and *building regulation* is life safety; this is why it is important to make sure that each building code is context specific. Building regulations seek to ensure that the policies set out in the relevant legislation are carried out. They contain the rules for building work in new and altered buildings to make them safe and accessible and limit environmental damage. Building codes, therefore, provide a good basis for incorporation of the DRR measures.

The *Eurocodes* provide structural design guidance for engineers in the design of the built environment. Related to disaster risk management, 'Eurocode 8 – Design of Structures for Earthquake Resistance' has particular significance. The objectives of Eurocode 8 in implementing earthquake resistance within structures is to protect human lives, limit damage, and to ensure important structures for civil protection remain operational. A key requirement of Eurocode 8 is to prevent collapse of structures. This involves designing to ensure that buildings have the capacity to withstand the seismic action without local or global collapse, and to be able to retain structural integrity and residual load bearing capacity after the event. While Eurocodes apply to all Member States, a national annex is present in Eurocodes to allow for individual nations to provide their own nationally determined parameters and/or additional information or rules.

Case Study 2.3

PreventionWeb

PreventionWeb is a participatory web platform for the DRR community. It was launched by UNISDR (see Case Study 2.1) with the purpose to facilitate an understanding of DRR by non-specialists. PreventionWeb provides a common platform for institutions to connect, exchange experiences and share information about DRR. Readers can submit content to PreventionWeb, as well as syndicate PreventionWeb content to their own websites. The website is updated daily and contains DRR news, events, online discussions, contact directories, policy documents, reference documents, training and academic programmes, jobs, terminology, and country information, as well as audio and video content.

Search terms: PreventionWeb; DRR database; UNISDR

Box 2.1
The earliest known written (if not rather draconian) building code is the Code of Hammurabi, in ancient Mesopotamia, dating back to 1772 BC; it states that *"If a builder builds a house for someone, and does not construct it properly, and the house which he built falls in and kills its owner, then that builder shall be put to death"*.

Standing Committee on Structural Safety (SCOSS) is an independent body established in 1976 and is supported by the Institution of Civil Engineers, the Institution of Structural Engineers and the Health and Safety Executive to maintain a continuing review of building and civil engineering matters affecting the safety of structures. The main purpose of SCOSS is to identify (in advance) trends which may contribute to an increased risk to structural safety, through risk analysis. SCOSS interacts with multiple stakeholders, including construction practitioners, industry and government on matters concerning the design, construction and operation of structures and buildings. SCOSS has introduced the Confidential Reporting on Structural Safety (CROSS) scheme. This allows for the collection and dissemination of information related to structural safety concern to provide lessons learnt, with the comments and data provided treated anonymously.

Table 2.2 Overview of the Disaster Risk Reduction Process.

Stage		Descriptor	
1	Identify, characterize, and assess hazards/threats	**Hazard/Threat identification** – the process of finding, recognising and describing hazards/threats to which the location is exposed.	
2	Assess the vulnerability of urban spaces to specific hazards/threats	**Vulnerability assessment** is the process of assessing the susceptibility of the intrinsic properties (the structure, materials, construction and planning) to a hazard/threat that can lead to an event with a consequence	
3	Determine the risk (i.e., the expected consequences of specific hazards/threats on specific assets)	**Identifying the level of risk** - magnitude of a risk or combination of risks, expressed in terms of the combination of the likelihood (chance of something happening) and the impact (consequences) of an incident caused by that hazard/threat. It utilises a **Risk Matrix** as a tool for ranking and displaying risks by defining ranges for consequence and likelihood	
4	Identify ways to reduce those risks	**Inherent safety** (eliminate) the possibility of hazards/threats occurring) **Prevention** (reduce the likelihood of hazards/threats) **Detection** (measures for early warning of hazards/threats) **Control** (limiting the size of the hazards/threats) **Mitigation and adaptation** (protection from the effects of hazards/threats) **Emergency response** (planning for evacuation and access for emergency services)	**Identifying (and prioritising) a course of action to address and treat the hazard/threat and its associated risks.** Treatment can involve: • avoiding the risk by deciding not to start or continue with the activity that gives rise to the risk; • removing the hazard/threat source; • changing the likelihood or magnitude; • changing the consequences; • protecting assets/spaces from the effects of the risk • preparedness planning for the impacts of risks (events) • sharing the risk with another party or parties [including contracts and risk financing]; and • retaining the risk by informed decision making
5	Prioritise risk reduction measures		

Source: After Bosher (2014) and British Standards Institution (2009)

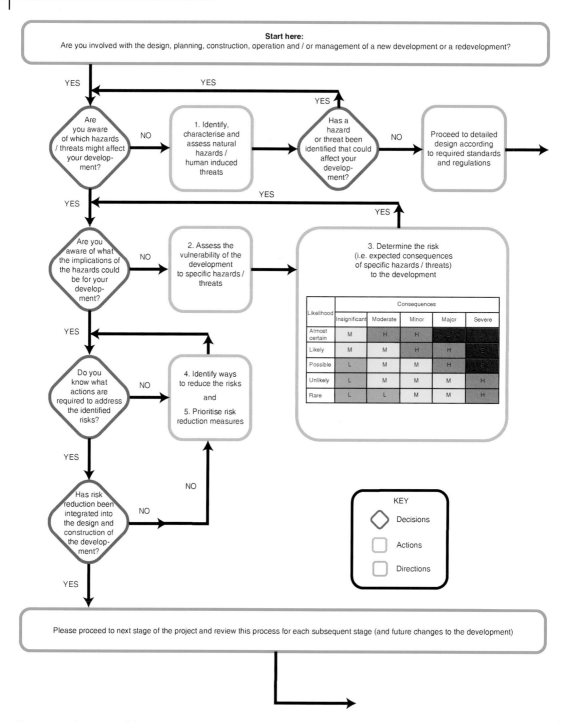

Figure 2.13 Overview of the risk management decision-making framework.

2.6   Summary

The profession of disaster risk reduction has been gradually developing since the mid-twentieth century. With the help of the UN, many risk reduction actions are now accepted across the globe and are increasingly being mainstreamed. Disaster risk reduction includes five stages, working through which can aid the consideration of prior knowledge of existing hazards and threats in a local context and to recognise that too often disasters occur because risk reduction measures have not been considered or undertaken even though there was knowledge of underlying hazards.

Table 2.2 provides a summary and Figure 2.13 provides a flow diagram of the disaster risk reduction framework which will be used as a basis for this book. The following chapters will be discussing various hazards and threats and risk reduction measures based on these stages.

Key points:

- Disaster risk reduction is a multi-disciplinary concept that requires the inputs of many stakeholders, over and beyond the traditional emergency management disciplines.
- There is now more recognition that we (collectively) need to change the way we develop our cities, infrastructure and buildings, by not merely mainstreaming DRR into practice but by making DRR part of the 'developmental DNA'.
- Current efforts in DRR are based on the Sendai Framework for Action in Disaster Risk Reduction that was agreed in 2015.
- Disaster risk reduction includes five stages: identification of hazards, assessment of vulnerabilities, determination of risks, identification of risk reduction measures, and their prioritisation.

Further Reading

Books:

Bicknell, J. (2009). *Adapting cities to climate change*. London: Earthscan.

Bosher, L (ed) (2008) *Hazards and the Built Environment: Attaining Built-in Resilience*, Taylor & Francis Group, ISBN: 9780415427296.

Mitchell-Wallace, K. Jones, M., Hillier, J. K., and Foote, M. (2017), *Natural catastrophe risk management and modelling: A practitioner's guide*, London: Wiley.

Pelling, M., & Wisner, B. (Eds.). (2008). *Disaster risk reduction: Cases from urban Africa*. Earthscan.

Vale, L. J., & Campanella, T. J. (2005). *The resilient city: How modern cities recover from disaster*: Oxford University Press, USA.

Wamsler C. (2013) *Cities, Disaster Risk and Adaptation*, Routledge

Wisner B., Blaikie P., Cannon T. and Davis I., (2004) *At Risk: Natural Hazards, People's Vulnerability, and Disasters*, second edition, Routledge, London.

Wisner, B., Gaillard, J. C., & Kelman, I. (Eds.). (2011). *Handbook of hazards and disaster risk reduction*. Routledge

Journal Publications:

Alexander, D. (2013). Resilience and disaster risk reduction: an etymological journey. *Nat. Hazards Earth Syst. Sci.*, 13, 2707–2716.

Bosher, L. (2014). Built-in resilience through disaster risk reduction: Operational issues, *Building Research and Information*, 42(2), pp. 240–254.

de la Poterie, A.T. and Baudoin, M.A. (2015). From Yokohama to Sendai: Approaches to Participation in International Disaster Risk Reduction Frameworks. *Int J Disaster Risk Sci*, 6, 128–139.

Jabareen, Y. (2012). Planning the resilient city: Concepts and strategies for coping with climate change and environmental risk. *Cities*, 31, 220–229.

Roberst, S. (2008). Effects of climate change on the built environment. *Energy Policy*, 36, 4552–4557.

Reports:

Alexander, D (2013). *Paradoxes and perceptions: Four essays on disasters*. IRDR Occasional paper.

Benson C. and Twigg J., (2007), *Tools for Mainstreaming Disaster Risk Reduction: Guidance Notes for Development Organisations*, International Federation of the Red Cross and Red Crescent Societies/ The ProVention Consortium, Geneva.

Birkmann, J. (2006). Measuring vulnerability to natural hazards: towards disaster resilient societies. Available at: http://archive.unu.edu/unupress/sample-chapters/1135-MeasuringVulnerabilityToNaturalHazards.pdf

British Standards Institution, (2009). *Risk management: Principles and guidelines*. London: British Standards Institution Group

CIB (2016), '*Disasters and the built environment: A Research Roadmap*', International Council for Research and Innovation in Building and Construction (CIB), Delft (Authored by Bosher L., Von Meding J., Johnson C., Farnaz Arefian F., Chmutina K. and Chang-Richards Y.) ISBN 978-908-0302-2-04

Prasad, N., Ranghieri, F., & Shah, F. (2009). *Climate resilient cities: a primer on reducing vulnerabilities to disasters: World Bank Publications.*

Rittinger et al., (2015). A new climate for peace. Available at: https://www.newclimateforpeace.org/

Twigg, J. 2009. Characteristics of a disaster-resilient community. Available at: http://practicalaction.org/ docs/ia1/characteristics-disaster-resilient-community-v2.pdf

World Economic Forum, 2015. Global Risks report. Available at: http://www.weforum.org/reports/ global-risks-report-2015

UN, (2015). Sendai Framework for Disaster Risk Reduction. Available at: http://www.preventionweb.net/ files/43291_sendaiframeworkfordrren.pdf

UNISDR, (2015). Global Assessment report on disaster risk reduction. Available at: http://www.unisdr. org/we/inform/gar

Other useful links:

Natural Hazards Centre https://hazards.colorado.edu/

International Disaster Database (EM-DAT): http://www.emdat.be/

Making cities resilient: http://www.unisdr.org/campaign/resilientcities/

PreventionWeb: http://www.preventionweb.net/english/

UN Habitat: http://unhabitat.org/

UNISDR: http://www.unisdr.org/

US Federal Emergency Management Agency: http://www.fema.gov/

Section II

Hydro-Meteorological Hazards

3

Flooding

"Rising sea levels will result in tens to hundreds of millions more people flooded each year with a warming of 3 or 4°C. There will be serious risks and increasing pressures for coastal protection in South East Asia (Bangladesh and Vietnam), small islands in the Caribbean and the Pacific, and large coastal cities, such as Tokyo, Shanghai, Hong Kong, Mumbai, Calcutta, Karachi, Buenos Aires, St. Petersburg, New York, Miami and London."
– Nicholas Stern, The Economics of Climate Change: The Stern Review, 2007

Water covers around 70% of Earth's surface and is essential for supporting life on the planet. From the beginning of civilisation, people have been settling near the water's edge, because the land adjacent to water has offered many advantages. However, water can also be a significant hazard due to the repeated threat of flooding. Flooding is the most prevalent natural hazard globally, and will remain as such as long as people live and work in flood-prone areas.

3.1   Learning Objectives

By the end of this chapter you will learn:

- What are the different types of flooding?
- What are the main causes of flooding?
- What are the typical impacts of flooding?
- How to identify the risk of flooding and assess vulnerabilities?
- How to reduce the risk of flooding using structural and non-structural measures?

3.2   Living with Water

Humankind has historically been attracted to the water's edge, such as the sea, rivers and lakes. People like to live near to water, mainly for the benefits of drinking, food, transport, energy and industry that such locations can provide in abundance. However, these locations tend to provide environments that are hazardous due to a broad range of flooding events related to unmanageable amounts of excess water.

Throughout the history of human development, floods have helped to bring untold wealth and prosperity to human habitations, and yet at the same time, they have also caused incredible losses and resulted in extensive suffering for millions of people. The most disastrous floods during the last

Disaster Risk Reduction for the Built Environment, First Edition. Lee Bosher and Ksenia Chmutina.
© 2017 John Wiley & Sons Ltd. Published 2017 by John Wiley & Sons Ltd.

century are listed in Table 3.1. Long-term statistics collated by the UNISDR (see Figure 3.1) indicate that since 1980, the amount of floods occurring each year are on the increase and that overall floods pose the greatest risks globally to people, essential lifeline services and economic stability and growth (UN GAR, 2013).

Table 3.1 List of the 10 Most Disastrous Floods in the Last Century.

Year	Event	Location	Death toll
1931	China floods	China	2.5-3.7m
1938	Yellow River (Huang He) flood	China	0.5-0.7m
1975	Banqiao Dam failure	China	231,000
1935	Yangtze River flood	China	145,000
1971	Hanoi and red River Delta flood	Vietnam	100,000
1949	Eastern Guatemala flood	Guatemala	40,000
1954	Yangtze River flood	China	30,000
1974	Bangladesh monsoon rain	Bangladesh	28,000
1999	Vargas mudslide	Venezuela	20,006
1939	Tianjin flood	China	20,000

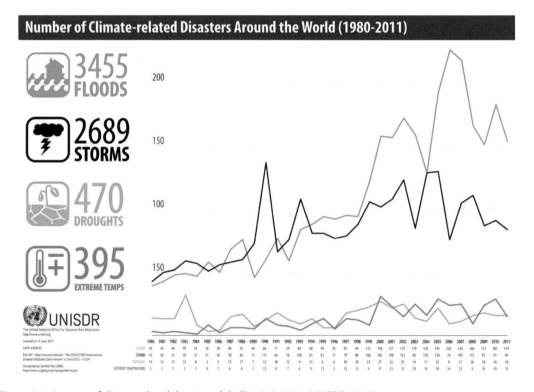

Figure 3.1 Amount of climate related disasters globally 1980–2011 (UNISDR, 2012).

Floods can occur in various ways (see Figure 3.2), and are typically categorised as:

- River (or fluvial) floods
- Coastal floods
- Flash floods
- Urban (pluvial) floods

It is important to categorise these floods because different types of floods can be caused by different things (natural and human-induced), may have varying levels of predictability and intensity and can cause different types of damage. Consequently, the risk management approaches required for these flood types will need to be different.

3.3   Overview of the Typical Impacts of Floods

According to detailed records compiled by EM-DAT, in the 50 years between 1964 and 2013 there were 4,000 major flood events globally, that killed nearly 300,000 people, affected 3.5 billion people and caused an estimated US$632bn in damages (Table 3.2).

The primary effects of flooding include loss of life, damage to buildings and other structures, including bridges, sewerage systems, roadways, and canals.

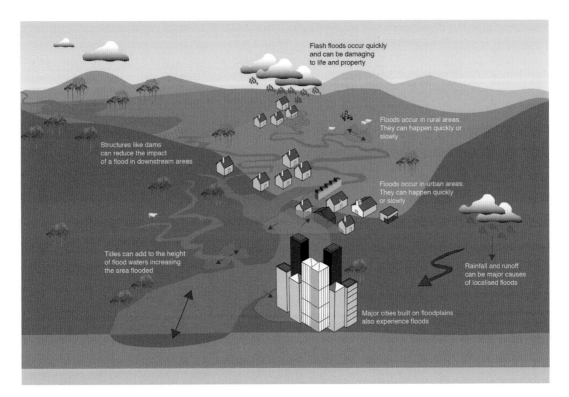

Figure 3.2 Illustration of the different types of flood risk (*Source:* WEDC).

Table 3.2 Flood Averages Per Year Between 1964 and 2013.

Flood incidents	83
Deaths	5,912
Total people affected	70,525,601
Total economic damages (US$)	12.7bn

Secondary effects such as economic hardship due to a temporary decline in business transactions, reduced revenues from tourism and rebuilding costs are common after a severe flood. The impact on those affected may cause psychological stress, in particular where deaths, serious injuries and loss of property occur. Extensive flooding can lead to chronically wet houses, which are linked to an increase in respiratory problems and other illnesses. Coastal flooding can result in the salinisation of agricultural land, whereas urban flooding invariably has significant economic implications for affected neighbourhoods (more examples of the primary and secondary effects can be found in Table 3.4).

3.4   Causes of Flooding

Disasters caused by floods are often referred to as 'natural disasters', implying that the main causes of the disaster are the natural hazards (see Thinking Point 1.1 in Chapter 1). In many cases this is indeed a reasonable comment, with factors such as intensity of rain/snowfall, size of the catchment, ground absorbency, topography and local vegetation (to name just a few) playing important roles. However, there is increasing acknowledgement that flood risk can be exacerbated, or even generated, by human activities; such as:

- Poor (or unregulated) urban planning (building homes, offices, infrastructure and essential services on flood prone land), see Figure 3.3.
- Buildings and infrastructure not designed to cope with the physical impacts of floods
- Increased urbanisation (leading to quicker run-off from surfaces to water courses)
- Lack of, or ineffective, urban drainage systems (to cope with expanding developments or impacts of climate change)
- Poorly conceived river/flood management schemes (leading to over reliance of physical assets that need long-term maintenance and may become quickly outdated)
- Deforestation of steep slopes (leading to instable surfaces and increased run-off)
- The construction of dams (leading to flooded tracts of land, they also pose flood risks due to the potential structural failures of dams).
- Lack of, or ineffective, emergency preparedness procedures.

All of these factors are human induced and play important roles in how natural processes such as river hydrology can turn into hazards that result in disasters. Importantly, these are the key factors that those involved in how the built environment is planned, designed, built, managed and upgraded, can influence by playing a more positive role in addressing disaster risk as an everyday part of their jobs.

Figure 3.3 Flooding in Carlisle, England in January 2005. Poor planning contributed to critical services such as the emergency services, local government offices and electricity substations being located in flood prone areas (*Source:* © ICDDS).

3.5   Riverine Floods

Riverine (or fluvial) floods – rainfall over an extended period and an extended area can cause major rivers to overflow their banks. Downstream areas may be affected, even when they didn't receive much rain themselves.

Key factors:

Main causes: Riverine floods are influenced by the hydrology of water courses, the size and topography of the catchment, the intensity and location of rainfall, and how the built environment has been designed.

Characteristics: River catchments (such as the Amazon, Ganges and Mississippi) can cover extensive areas; therefore, heavy rainfall can occur on high ground within the catchment but a long distance from the river, and this can result in a time lag between the rain event and any flooding downstream.

Key impacts: Riverine floods can result in floods that take days (or even weeks) to subside, but effective warnings can help to minimise the loss of lives. Buildings and critical infrastructure can be damaged or destroyed and it can become extremely difficult for affected areas to function for a prolonged period; economic losses can be vast. Chemicals and other hazardous substances (such as sewage) can contaminate the water bodies and people's homes.

Location: We know where riverine floods can strike as they tend to have been flooded in the (recent) past. Studying a good quality map with contour lines (such as the Ordnance Survey maps in the

NORMAL CONDITIONS   FLOOD CONDITIONS

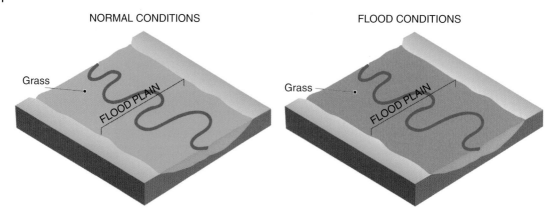

Figure 3.4 A flood plain is an area of land adjacent to a stream or river that stretches from the banks of its channel to the base of the enclosing valley walls and experiences flooding during periods of high discharge (*Source:* WEDC).

UK) will give a quick indication of what areas within a river catchment are prone to flooding; particularly in high risk areas such as the natural flood plain (see Figure 3.4).

Advance warnings: Because we know where rivers and their associated flood plains are located, it is possible to consider this information when planning new developments. Although this may sound obvious, it is not always the case that flood prone locations are avoided when building new developments (see Thinking Point 3.1). Also, the potential time lag on large catchments between rainfall and when the river levels rise indicates that flood warnings can be relayed to any areas at risk of floods.

3.6   Coastal Floods

Coastal floods – a coastal flood is when the coast is flooded by the sea. These floods can be caused by surges associated with a severe storm, and exacerbated by sea level rise. Storm driven winds can create high waves.

Key factors:

Main causes: Coastal floods are influenced by the topography and geology of the coastline, the heights of the tidal ranges, and susceptibility of the area to coastal storms, impacts of localised sea level rise, types of land use and how the built environment (including coastal erosion/flood defences) has been designed.

Characteristics: Coastal areas that include large areas of low-lying land or river deltas (such as the Pearl River delta in China, the Mississippi delta in the USA and most of the coast line of Bangladesh) can cover extensive areas and are prone to coastal floods. Low pressure storms (such as cyclones, see Chapter 4) can generate coastal storm surges that can travel large distances inland, particularly if coinciding with high tides; the likelihood and magnitude of these events are exacerbated by sea level rise.

Thinking Point 3.1

Carlisle Flood – What Went Wrong and Why?

Carlisle is situated on the flood plain of the River Eden with three rivers meeting in the city. Carlisle has already had a history of flooding with the most recent flooding occurred in January 2005, peaking on 8 January. Heavy rainfall across Cumbria led to significant flooding in the city of Carlisle, which left three people dead, 1,844 properties flooded and led to significant disruption to residents, businesses and visitors. The cost of the flooding was estimated at over £400 million.

The extreme nature of the weather and flooding meant that the Environment Agency's flood models were not very accurate and did not predict the magnitude of the flood event well. Fifty percent of residents in the flood risk area had not signed up to receive a warning. Carlisle was cut off and awash for some time on 8 January, with flood waters nearing the second floors of a few houses, leaving residents stranded. Some had to be airlifted off the roofs of their homes. Evacuation was dangerous in darkness and high winds.

Carlisle's existing flood defences were built after flooding in 1968, however, the old flood defences were over-topped, resulting in a sudden, dangerous water rise. As the cities flood defences failed, problems were compounded by widespread transport disruption; 2.5 metres of water flooded the fire station and police station adding to the problem of help by these services. It appeared that Police and Fire services' headquarters were located in one of the most severely flooded areas. Not only did this significantly reduce the effectiveness of emergency response, but there were reports of looting, arson and lawlessness within the town as a result. The areas critical infrastructure, including police, fire and rescue services and telecommunications facilities, received no greater protection than other parts of the city.

The Carlisle example illustrates a failure to prioritise risk reduction measures that protect critical infrastructure. It is telling that whilst Carlisle has since upgraded its flood defences, police and fire services have since been moved out of the town. Worse, this lack of prioritisation led to dire consequences as the emergency response was badly hampered. Carlisle also highlights a weakness in urban planning, by locating essential facilities in areas vulnerable to flood.

As a result of the 2005 flood event, the Environment Agency produced a flood alleviation scheme to reduce the current and future risk of flooding. Climate change is predicted to result in changes to flooding, with the Eden catchment indicating that winter rainfall (and therefore flood flows) might increase by as much as 15% to 20% by 2050, leading to increased flooding in urban areas.

Useful search terms:

Carlisle floods, Cumbria fire and rescue, emergency planning, River Eden, Environment Agency

---

Key impacts: Coastal floods can result in high impact floods that can eat into coast lines and wash away buildings and infrastructure, but effective warnings and evacuation plans can help to minimise the loss of lives. Buildings and critical infrastructure can be damaged or destroyed and it can become extremely difficult for affected areas to function for a prolonged period; economic losses can be vast if critical infrastructure (such as roads, railways and energy generate and distribution networks are damaged).

Location: We have a good insight into where coastal floods can strike as they tend to have been flooded in the past, or there is evidence of coastal erosion/flood defences being built. However, the impacts of sea level risk suggest that new locations may be affected in the future; therefore, local risk assessments should help to identify areas/assets that are most at risk of being affected (see Case Study 3.1).

CASE STUDY 3.1

Winter UK Floods 2013-2014

In December 2013 severe storms hit the UK with further heavy rain and strong winds continuing throughout the Christmas period and into the New Year. A large tidal surge, high tides and strong winds exposed large parts of the UK to flood risk not only on the coast but also inland flooding from swollen rivers and surface water, especially in the South, South West and East of England, adding to ongoing risks of tidal flooding to the South and West coasts of England as well as coastal parts of Wales. A major focus of concern was high spring tides and large waves combining to cause an extreme risk of coastal flooding. The historic promenade in Aberystwyth was severely damaged by large waves, and there was further damage and flooding to coastlines elsewhere, particularly the South Coast and Welsh coastline.

Over 2,800 km of flood walls, banks and other flood defences along the English coast and estuaries protected more than 800,000 properties from flooding (EA, 2014), however about 750,000 properties were left without power supply, and 1,400 houses were flooded on the east coast of the UK. It is worth noting that around 100 metres of sea wall in the town of Dawlish was destroyed by storms in early 2014 causing a significant stretch of railway to collapse into the sea (see Figure 3.5). Since the railway line was originally constructed in 1845, regular repairs and upgrades have been required to the sea wall due to relentless battering of the artificial defences by the sea.

Useful search terms:

Dawlish rail line, UK Winter storms 2013, MetOffice warnings, Network Rail

---

Figure 3.5 The stretch of track at Dawlish in south Devon was left hanging when the sea wall built to protect it was destroyed during a storm, which battered the south west of England in February 2014 (*Source:* Courtesy of Sarah Ransom).

Advance warnings: Advances in weather forecasting and the use of satellite imagery now make it possible for relevant authorities and the general public to be given one or two days advance warnings of coastal flood risk. While effective warnings and evacuation plans can help to minimise the loss of lives, they will do little to protect vulnerable buildings or infrastructure.

3.7 Flash Floods

Flash floods – can be difficult to predict as they are a very direct response to rainfall with a very high intensity or sudden massive melting of snow. The area covered by water in a flash flood is relatively small compared to other types of floods.

Key factors:

Main causes: Flash floods are influenced by the hydrology of water courses, the size and topography of the catchment, the intensity and location of rainfall, and how the built environment has been designed and maintained.

Characteristics: Areas that are prone to flash floods are typically quite small and affected by heavy rainfall localised rainfall. Rain water collects on the slopes of a valley and quickly accumulates in the water channel that can swiftly swell in size generating fast-moving flood waters.

Key impacts: Flash floods can result in floods that arrive with little or no warning and then subside just as quickly. The difficulty in predicting the location of such floods means that effective warnings are unlikely. The speed of the flood waters means that buildings and critical infrastructure can be damaged or destroyed (see Case Study 3.2). It can become extremely difficult for affected areas to function for a prolonged period; economic losses can be vast. The affected area can be left littered with large amounts of debris such as cars and rocks that have been washed down the river channels.

CASE STUDY 3.2

Boscastle

On the 16 of August 2004 the village of Boscastle in Cornwall experienced flash flooding when an exceptional amount thundery showers fell over eight hours (with 60 mmm – the month's average rainfall – falling within two hours). Boscastle is located at the confluence of three rivers – Valency, Jordan, and Paradise; there was simply too much water for the small rivers and the surrounding hillside to cope with, and the water gushed rapidly down the steep slopes into Boscastle. In addition, the flooding coincided with a high tide, making the impact worse. The flood was exacerbated by the fact that the ground was already saturated due to the previous two weeks of above average rainfall.

Due to the rapid response of the emergency services, no lives were lost, however homes, businesses and cars belonging to more than 1,000 people were swept away, leading to a vast number of subsequent insurance claims. Being a popular tourist destination, Boscastle economy was also affected by the flood, as a number of tourists have reduced dramatically. The damages were caused mainly because of a lack of flood protection and old, small drainage systems which were insufficient to cope with the flow; and trees and vehicles blocked the river channel, particularly under a bridge in the centre of Boscastle, which led to the river bursting its banks and further flooding in nearby properties.

Useful search terms: Boscastle; 2004 flash flood; river flooding

Location: It is very difficult to predict where flash floods will strike. However, looking at the characteristics of past flash flood events can give some indication of what areas within a river catchment are prone to this type of flooding.

Advance warnings: Flash floods are very difficult to predict, due the lack of any lag time between the rainfall event and the almost immediate impacts on water levels in the local rivers and streams.

3.8   Urban (Pluvial) Floods

Urban or pluvial floods – can be caused by flash floods, coastal floods, or river floods, but there is also a specific flood type that is called urban flooding. Urban flooding is specific in the fact that the cause is a lack of drainage in an urban area.

Key factors:

Main causes: Urban floods are influenced by the design and capacity of the urban drainage systems, the size and topography of the city, the intensity and location of rainfall and generally how the built environment has been designed.

Characteristics: Cities (such as Los Angeles, Mexico City, Kolkata and Bangkok) can cover extensive areas of hardened surfaces that have effectively reduced the absorptive capacity of the land. Roads, pavements and roofs have replaced land that was once effective at absorbing rain but now hardened surfaces (particularly in tropical climates where rain can be intensive) quickly channel the water into drainage channels that, if such infrastructure exists, can quickly become overburdened.

Key impacts: Urban floods can result in floods that arrive with little or no warning and then may take days to subside. The difficulty in predicting the location of such floods means that effective warnings are unlikely. The speed of the flood waters means that buildings and critical infrastructure can be damaged or destroyed. It can become extremely difficult for affected areas to function for a prolonged period; economic losses can be vast. The affected area can be left flooded for a prolonged period, particularly if drainage channels have become blocked or silted up (see Case Study 3.3).

Location: We can get an insight into where urban floods can strike as such locations are either low lying and/or without suitable drainage systems. Studying a good quality map with contour lines and speaking to local residents will give a quick indication of what areas within a city are prone to urban flooding. It should be possible for areas within a city to be identified as vulnerable to urban floods, thus it is possible to consider this information when planning new developments.

Advance warnings: Urban floods can be difficult to predict, due the lack of lag time between the rainfall event and the quick impacts on water levels in the city.

3.9   Risk Management

3.9.1   Historical Approaches

In the past societies have tended to be aware of their natural environment: they planned their settlements and built their homes and infrastructure in a way that minimised any disruptions from natural hazards. In settlements located near to rivers or the coast, important buildings were typically built on higher ground (see Figure 3.6) that was less prone to flooding.

CASE STUDY 3.3

Urban Floods in Bengaluru, India

Bengaluru (previously known as Bangalore), and many other cities worldwide, has been severely affected by the impacts of climate change – flooding in particular. Factors such as urban ecological degradation, inadequate drainage and informal settlements, mean that the risks of urban flooding have increased. The mean annual rainfall in Bengaluru is about 880 mm with about 60 rainy days a year. It is the fifth largest city of India with a population of about 7 million, located around 100 km from the Kaveri River. Bengaluru has seen intense growth of its population (632% from 1973 to 2009) alongside slum development; according to the 2011 census there were over 500 slum settlements in greater Bengaluru. Not surprisingly, the encroachment of informal settlements near water bodies and an increased built-up area of housing, office parks and other infrastructure (amounting to 23% of the city's land, up from 16% in 2000) has resulted in the decline of water bodies in the metropolitan region by 58%. Encroachment of wetlands and floodplains is causing floodway obstruction and loss of natural flood storage in Bengaluru. In addition, lake catchments were being used as dumping yards for municipal solid waste, construction residue or building debris. Bengaluru city has a 180-km-long primary and secondary storm-water drainage system, which also often fails to take the load of the rains because silt and garbage cause blockages. All these reduce Bengaluru's capacity to cope with rain. In 2005, because of the unauthorised developments among three lakes, blocked drains led to residential areas being inundated and traffic severely affected. Thousands of people were stranded on the city's water-logged roads, schools were closed and several apartment blocks and office buildings (including WIPRO, one of the India's largest software companies) were flooded.

A provision of Rs 45 million (approximately £450,000) has been made for the flood-management fund with 12 squads on call, of which six are rain and flood relief squads; 20 personnel have been assigned in each squad. The Jawaharlal Nehru Urban Renewal Mission (JNURM) project was launched in December 2005 and Bengaluru has been allocated a budget to enhance its flood resistance. Clearly urban planning strategies have to be changed dramatically because they are affecting social and economic development of the city and the country.

Useful terms: flood management; JNURM: Bengaluru, floods

Generally, people have been effective at adapting to the hazards posed by floods, mainly because flood-prone areas were good for fishing, agricultural use and transportation. Some of these adaptations have included:

- building complex systems to control river flows (i.e., the Shushtar hydraulic system built in Iran during the fifth century BCE) (see Figure 3.7)
- establishing flood warning systems (i.e., by the Yuan Dynasty in China between 1279-1368)
- building houses elevated above the water (see Figure 3.8)
- using materials (i.e., lime render) that are more resilient to the impacts of floods.

However, since the industrial revolution that started in the eighteenth century, cities have flourished in economic importance and population growth has resulted in situations where rapid urban development has been unregulated, poorly planned and inadequately built with little regard for traditional approaches to dealing with flood risk. We now are living with the legacies of these 'poor decisions' but sadly in many parts of the world these 'poor decisions' are still being made today.

Figure 3.6 Tewkesbury Abbey located on high ground and thus protected from the 2007 floods that affected the rest of the Tewkesbury, England (*Source:* ICDDS).

Figure 3.7 Shushtar, Historical Hydraulic System (in modern day Iran) inscribed as a UNESCO heritage site is an ancient wonder of water management that can be traced back to Darius the Great in the fifth century B.C. (*Source:* Reproduced with permission of Hadi Nikkhah).

Figure 3.8 Traditional house elevated over a floodplain in Cambodia (*Source:* Reproduced with permission of Southopia on Flickr).

3.10 Hazard Identification

There are a number of ways to identify the hazard of flooding. Some national governments and local authorities produce (or at least make available) flood risk maps in a number of scales and resolutions. In England, the Environment Agency (EA) has produced a searchable flood risk map (see Figure 3.9

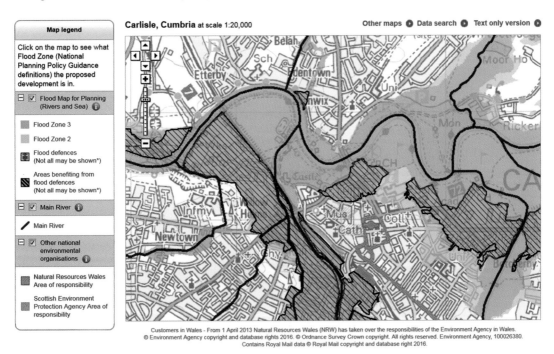

Figure 3.9 Example of an EA flood risk map that has been designed for use by the general public, in this case showing Carlisle, England (*Source:* Environment Agency (EA)).

and Case Study 3.4) that provides useful information about flood risk (including impact of flood defence measures). Insurance companies tend to be proactive in producing flood risk maps but very often these maps can be considered commercially sensitive and thus may not be available to the general public or even made accessible to other key decisions makers.

There are a number of scientific and non-scientific ways in which flood-related hazards can be identified and assessed. The most prominent methods are summarised in Table 3.3 for the four main types of flood risk.

CASE STUDY 3.4

Flood Maps

The role of the flood maps in flood risk management is on the increase. They are used not only for defining, illustrating and communicating flood risks, but also for regulating them and for recognising the limitations of the proposed flood mitigation measures.

In England, publically available online-based flood maps are produced by the Environment Agency (see Figure 3.9). They depict the probability of flooding, using different colours to mark out zones exposed to different levels of risk from fluvial and tidal flooding. However, the aim of these maps is to show the risks as well as to restrict development exposed to flooding – but doing so indirectly by influencing and managing the activities of third parties (see Thinking Point 3.1), in particular the decision making of local planning authorities that are responsible for granting development permission, rather than through any outright prohibition.

Useful search terms:

Environment Agency, flood risk maps, local government GIS, flood return periods

---

Table 3.3 Overview of Typical Flood-Related Hazard Identification Methods.

Type of flood	Hazard identification methods
River (or fluvial) floods	We may know where riverine floods can strike because certain locations may have been flooded in the (recent) past and maybe a record kept of flood heights (see Figure 3.10 of flood marker in Cockermouth). Although this suggestion may sound like common sense, it is common sense that is not always applied. Studying a good quality map with contour lines will give a quick indication of what areas within a river catchment are prone to flooding; particularly in high risk areas such as the natural flood plain.
Coastal floods	We have a good insight into where coastal floods can strike as they tend to have been flooded in the past or there is evidence of coastal erosion/flood defences being built. However, the impacts of sea level risk suggest that new locations may be affected in the future; therefore, local risk assessments and general local knowledge should help to identify areas/assets that are most at risk of being affected (see Case Study 3.2).
Flash floods	It is very difficult to predict where flash floods can strike. However, looking at the characteristics of past flash flood events can give some indication of what areas within a river catchment are prone to this type of flooding. Bear in mind that flash flood risk can be generated or exacerbated by inappropriate development and poor management and maintenance of existing water courses.
Urban (pluvial) floods	We can get an insight into where urban floods can strike as such locations are either low lying and/or without suitable drainage systems. Studying a good quality map with contour lines and speaking to local residents will give a quick indication of what areas within a city are prone to urban flooding. It should be possible for areas within a city to be identified as vulnerable to urban floods, thus it is possible for us to consider this information when planning new developments.

Figure 3.10 Photo of flood marker (height of the line is approximately 1.8 meters from pavement level) recording the November 2009 flood level in Main Street, Cockermouth, England (*Source:* Bosher, 2014).

3.11 Assessment of the Vulnerabilities

Once an understanding has been obtained of what the hazards might be, it is then critical to gain an appreciation of what types of sites, buildings, facilities and services could be affected by the identified hazards. This involves an assessment of the potential vulnerabilities of specific aspects of such locations that can be categorised as follows (and previously detailed in Chapter 2):

1) Physical vulnerability
2) Social vulnerability
3) Economic vulnerability
4) Environmental vulnerability

5) Governance vulnerability

Physical Vulnerability may be determined by aspects such as population density levels, the location and gradient of the site, design and materials used for critical infrastructure and for housing.
Example: Wooden homes are less likely to collapse in an earthquake, but they can be more vulnerable to the impacts of flash floods and coastal floods.

Social Vulnerability refers to the root causes that make it harder for some people (i.e., women, the elderly, ethnic groups) to withstand adverse impacts of flood-related hazards.
Example: Some people have limited choices about where to live, especially if they have migrated to large cities, and may end up inhabiting hazard prone locations (i.e., flood plains); thus, when flooding occurs some citizens, particularly children, women, elderly and the less mobile, may be unable to protect themselves or evacuate if necessary.

Economic Vulnerability. Levels of vulnerability can be highly dependent upon the economic status of individuals, communities and nations.
Example: Families that have migrated from rural areas to live in large cities such as Dhaka in Bangladesh may end up living in squatter settlements located on the flood prone banks of the Buriganga River because they cannot afford to live in safer (more expensive) areas. Such families may live hand-to-mouth existences and struggle to save funds that could help them in times of crisis (such as floods and storms but also more common/everyday crises such as ill health and unemployment).

Environmental Vulnerability. Natural resource depletion and resource degradation are key aspects of environmental vulnerability.
Example: Mangrove swamps along the coast of India have historically afforded some protection for coastal areas from storms and coastal floods. However in recent years many mangrove swamps have become denuded due to locals using the mangrove trees for firewood (see Figure 3.11) but also on a larger scale, coastal mangroves have been destroyed to made 'aquaculture farms' where shrimps/prawns can be bred in large man-made tanks containing saline water.

Governance Vulnerability. The complex roles of national and local government in producing disaster risk and also in being affected by those risks can too often become over looked.
Example: 'Thinking Point 3.1 Carlisle floods: what went wrong and why?' is a prime example of what can happened when local support mechanisms (i.e., police, social services, healthcare, fire and rescue emergency response and electricity distribution) become affected by floods and thus severely inhibit the affected population to cope and recover from such events.

If the proposed development (or use of the development) is inappropriate for the site, an alternative site should be sought or an alternative use considered. If no alternative sites are available, 'exceptional' reasons why the development use should be considered must be demonstrated (in England this is known as the 'Exception Test').

3.11.1   Appropriate Uses

Understanding the vulnerability of locations, sites, proposed buildings and an appreciation of the final use of such developments are essential; thus, local level assessments are particularly important when obtaining this information. Figure 3.12 provides an overview of the types of uses (buildings and types of activity) that could be appropriate, dependent upon the levels of identified flood risk. Figure 3.13 illustrates how flood zoning (based upon predicted flood return periods) can be used to encourage the siting of sensitive or critical services (i.e., houses and emergency services) in low flood risk areas and also limit developments in high flood risk to less sensitive recreational activities.

Figure 3.11 Last but vulnerable remnant of mangrove forest (that used to line this coastal area) on the east coast of Havelock Island, The Andaman and Nicobar Islands, India (*Source:* Bosher, 2003).

3.12 Determination of the Risk

Effects associated with flooding can be divided into primary and secondary. The primary effects of flooding are those caused by actual contact with flood water, such as death due to damage to buildings, land erosion, crop loss, and so on. The secondary effects are longer-term effects that are indirectly related to flooding (see Table 3.4 for more examples). Scientists have developed ways to calculate risks, on the basis of, for example, hydrological/river height statistics, meteorological statistics, climate change model predications, strength of current defences, estimates of the amount of affected population and economic loss estimates.

3.12.1 Flood Damage Estimation

Flood damage estimation can be a very useful exercise in the management of flood risk by informing the decision making process. Damage estimation is also used extensively in the insurance industry, in order to calculate the financial exposure an organisation or development may have to a flood event. When coupled with an inundation model, it is possible to understand the effects of flood defences on the amount of damage caused by a range of flooding scenarios. Figure 3.14 shows results of inundation modelling using different buildings treatments. The convention is to use a Digital Terrain Model (DTM), which strips the topography of all vegetation and other structures which may impede flow of water across the floodplain.

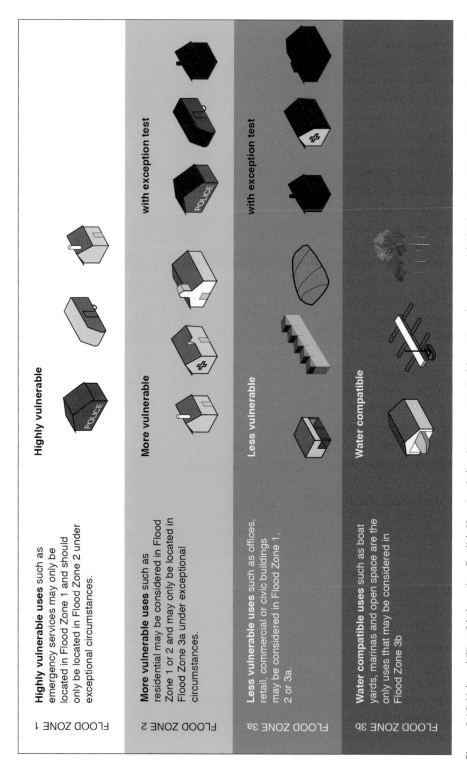

Figure 3.12 Vulnerability table based on English Planning Policy Statement 25. Buildings/sites that are used for high vulnerability activities (such as schools and emergency services) should ideally be located away from high flood risk areas (*Source:* WEDC at Loughborough University after LIFE 2013).

FLOOD ZONE 1

Highly vulnerable

Highly vulnerable uses such as emergency services may only be located in Flood Zone 1 and should only be located in Flood Zone 2 under exceptional circumstances.

FLOOD ZONE 2

More vulnerable

with exception test

More vulnerable uses such as residential may be considered in Flood Zone 1 or 2 and may only be located in Flood Zone 3a under exceptional circumstances.

FLOOD ZONE 3a

Less vulnerable

with exception test

Less vulnerable uses such as offices, retail, commercial or civic buildings may be considered in Flood Zone 1, 2 or 3a.

FLOOD ZONE 3b

Water compatible

Water compatible uses such as boat yards, marinas and open space are the only uses that may be considered in Flood Zone 3b.

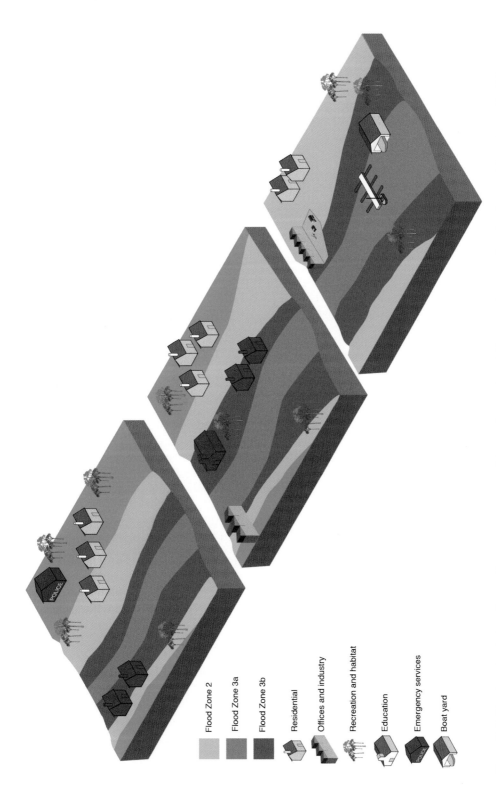

Figure 3.13 Flood zoning that can be used to encourage appropriate developments and discourage inappropriate developments based upon English Planning Policy Statement 25 (*Source*: WEDC at Loughborough University after LIFE 2013).

Flood Zone 2

Flood Zone 3a

Flood Zone 3b

Residential

Offices and industry

Recreation and habitat

Education

Emergency services

Boat yard

Table 3.4 Examples of the Primary and Secondary Effects of Flooding.

	Primary effects	Secondary effects
Physical	*Damage to...* • Critical infrastructure • Houses • Offices/shops • Schools • Vehicles • Equipment • Agriculture	*Resulting in....* • Homelessness/displacement • Disrupted essential services • Sedimentation of water channels/sewers • Disrupted education provision • Polluted land • Loss of crops
Wellbeing/health	*Deaths or injuries to...* • People • Livestock	*Leading to...* • Epidemic diseases (i.e., cholera) • Post-traumatic stress disorder • Unhygienic conditions • Polluted land/homes • Strains upon social support services
Economic	*Immediate...* • Repair/replacement costs • Loss of business	*Longer term...* • Rehabilitation costs • Insurance costs (and increased premiums) • Reduced commerce • Loss of tourist revenues • Reduced land values

3.13   Identification and Prioritisation of Risk Reduction Options

As discussed in Chapter 2, there are a number of ways to reduce the risk of a hazard. In case of flooding a number of options are available and these will be discussed further, however, it is important to bear in mind that these options should not be considered a panacea as they cannot always to provide a 100% guarantee of safety.

Fundamentally, when it comes to dealing with flood risk, the approaches can be summarised into the following three classifications (see Figure 3.15), adapted from ICE (2010):

1) **Retreat** – take a managed retreat approach, where it is acknowledged that long-term investment in artificial defences is not sustainable and the sea/river is allowed to reclaim the land. This approach may not be popular with all stakeholders as it could involve loosing land/property to sea/rivers. New investment must also be made in relocating communities and infrastructure out of harm's way. However, money is saved by significantly decreased investment in flood defences.

2) **Defend** – use existing defences or build more to protect current developments. By choosing to defend, the existing built infrastructure of a city is protected from floods and does not need to be relocated to higher ground or rebuilt after flooding. However, flood defence is an extremely costly endeavour.

3) **Attack** - To attack is to advance and step sea/riverward of the existing high waterline. There is massive development potential to be gained for cities by building out onto the water. This further reduces the need to sprawl into the countryside and ensures their sustained social and economic vitality. However, like flood defences, 'attack' is an extremely costly endeavour and there are many examples that show that such ('man over nature') approaches can prove to be unsustainable in the long-term.

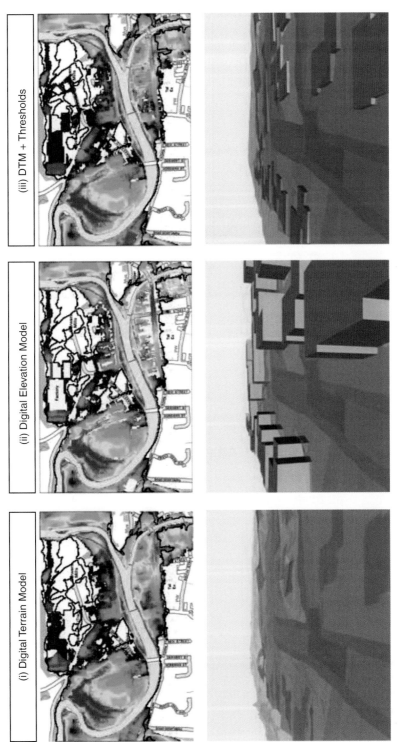

Figure 3.14 Inundation modelling results showing different building treatments. (i) Digital Terrain Model (DTM) treatment where all buildings and vegetation are stripped from the topography, (ii) DTM has the building footprints stamped on (iii) DTM+ Thresholds has buildings footprints stamped on at threshold height (*Source: O'Neill et al., 2011*).

(i) Digital Terrain Model

(ii) Digital Elevation Model

(iii) DTM + Thresholds

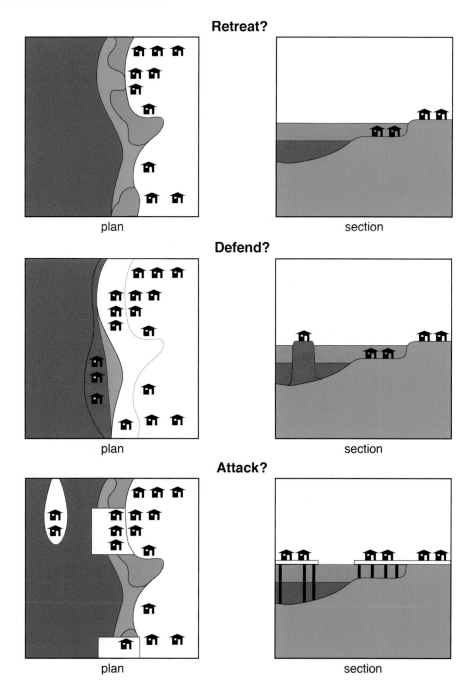

Figure 3.15 Three fundamental approaches to dealing with flood risk. a) Retreat - To retreat is to step back from the problem and avoid a potentially catastrophic blow. It is to move critical infrastructure and housing to safer ground and to allow the water into the city to alleviate flood risk, b) Defend - To defend is to ensure the sea water does not enter the existing built environment and c) Attack - To attack is to advance and step seaward of the existing coastline (*Source:* WEDC at Loughborough University after ICE 2010).

Table 3.5 Summary of the Viability of Risk Reduction Options for Addressing Flood Risk.

Type of risk reduction option for flood risk					
Inherent safety	Prevention	Detection	Control	Mitigation	Emergency response
#	#	Y	#	Y	Y

Please note:
'Y'– indicates that there are possibly a range of useful risk options available
'#'– indicates that some risk reduction options can be used but they are likely to be of only limited effectiveness
'N'– indicates that other than relocating the built asset there is little that can be done to reduce this hazard/threat

As outlined in Chapter 2, the best approach to considering risk reduction measures should adopt the five interrelated stages as shown in Table 3.5, which summarises the extent to which each of the risk reduction options can be utilised. A more detailed list of specific risk reduction examples is provided for dealing with flood risk in Table 3.6.

Table 3.6 Indicative Examples of Risk Reduction Options for Addressing Flood Risk.

Risk reduction option	Examples
1) Inherent safety – eliminate the possibility of hazards occurring	Do not build within the flood plain or on land adjacent to flood risk/coastal areas; Avoid redirecting/re-engineering the course of natural water courses (i.e., Mississippi River in USA)
2) Prevention – reduce the likelihood of hazards	*As above plus.* If environmentally appropriate build dams to moderate fluctuations in river flows; Managed retreat of development from flood threatened areas; Encourage natural barriers to grow/thrive (wetlands and mangroves); Use urban drainage systems of appropriate capacity
3) Detection – measures for early warning of hazards	Weather forecasting (using satellite imagery) – but not 100% reliable; River level gauges (with automated alarms) – expensive; Local knowledge and flood warnings from Environment Agency (or similar); Education and awareness raising of flood risk (in local schools and communities)
4) Control – limiting the size of the hazards	Protect natural river levees, flood plains and coastal margins; Artificial flood defences/walls/gates/ditches (i.e., Netherlands and LA); Sustainable urban drainage systems, swales and retention ponds; Land use zoning
5) Mitigation and adaptation – protection from the effects of hazards	Localised flood defences/walls/gates; Flood resilience/resistant materials for buildings and infrastructure; Flood-resistant defences for buildings and infrastructure; Amphibious buildings (or buildings/roads on stilts); Land use zoning; Developing an outreach program about flood risk and mitigation activities in homes, schools, and businesses; Car parking on ground floor; Non-return valves on sewer pipes;

(Continued)

Table 3.6 (Continued)

Risk reduction option	Examples
6) Emergency response – planning for evacuation emergency access	Land use zoning; Emergency planning and 'Local resilience forums'; Flood Liaison and Advisory Groups (e.g., in Scotland); Evacuation route planning; Locating critical infrastructure and emergency services/resources in safe/protected areas;

3.13.1 Prevention of Hazard

Dams can be used as a preventative measure against floods. However, while these typically expensive construction projects can help to reduce flood risk, they can also lead to a range of negative impacts such as relocation of local communities, loss of agricultural land, and loss of biodiversity (see Thinking Point 3.2).

Managed retreat of development from flood threatened areas can also be an option. Managed retreat (or managed realignment) allows an area that has previously been protected from flooding by the sea to become flooded by removing coastal protection. This process is usually adopted in low-lying estuarine areas where the long-term investment in artificial coastal defences has become unsustainable (see Figure 3.16 for an example). Managed retreat almost always involves flooding of land that has at some point in the past been claimed from the sea.

Encourage natural barriers to grow/thrive (mangroves and wetlands): Mangroves may help to mitigate disaster risk and damage in three ways, complementing other measures in a coastal defence strategy (see Figure 3.17). First, mangroves can reduce storm surge levels by up to half a meter for each kilometre of mangrove that the storm surge passes through. Secondly, the height of wind and swell waves is reduced by 13% to 66% within the first 100m of mangroves. These waves can be superimposed on top of storm surges, so their reduction can make a critical difference to storm impacts. Even during 'everyday' conditions, when such waves may be small, they still contribute to coastal erosion if their energy is not dissipated by a dense tangle of mangrove roots and branches. Thirdly, mangroves can help to stabilise sediments, both through their active growth and deposition of organic matter and by capturing sediments.

3.13.2 Detection of Hazard

Weather forecasting includes assessment of multiple sources of weather forecasting information and data and typically will include reviews of some or all of the following components:

a) *Recent weather conditions* – has there been a recent period of prolonged rain or other high impact weather?
b) *Rainfall forecasts* – is the forecast rain expected to be localised, short duration and high intensity or more widespread with a longer duration and of lesser intensity?
c) *Knowledge of catchment conditions*– how saturated are the catchments, how high are the rivers and what are the underlying conditions?
d) *Detailed flood forecast models for the coast*, showing surges and large waves, and flood flows for rivers are evaluated;
e) *Seasonal factors*, for example snow cover or leaf fall, and
f) *The combined effect of river flow and high tides* – if a river flood is being assessed, does this coincide with high tides, which could cause problems?

Thinking Point 3.2

Controversy of Dams

Dams are an important resource in many countries, providing various functions, including recreation, inland navigation, irrigation, water supply and hydroelectric power. Dams range from massive concrete structures to smaller scale earthen structures. Dams are also used for flood control: they can be effectively used to regulate river levels and flooding downstream of the dam by temporarily storing the flood volume and releasing it later. The most effective method of flood control is accomplished by an integrated water management plan for regulating the storage and discharges of each of the main dams located in a river basin. Each dam is operated by a specific water control plan for routing floods through the basin without damage. This means lowering of the reservoir level to create more storage before the rainy season.

However – despite their benefits – there is a big controversy around dams, with the main focus on their technical performance (as dams can fail with little warning). Intense storms and rainfalls may produce a flood in a few hours or even minutes for upstream locations. Flash floods can occur within six hours of the beginning of heavy rainfall, and dam failure may occur within hours of the first signs of breaching. Other failures can take much longer to occur, from days to weeks, as a result of debris jams, the accumulation of melting snow, build-up of water pressure on a dam with (unknown) deficiencies after days of heavy rain, and so on. Flooding can also occur when a dam operator releases excess water downstream to relieve pressure from the dam. The controversy is exacerbated by the social and environmental impacts dams may cause, with the Three Gorges Dam in China being a prime example:

Environmental impact: One of the most immediate environmental effects of the Three Gorges Dam has been an increase in landslide activity. This results primarily from erosion caused by the drastic increases and decreases in reservoir water levels. Another major issue is the ways in which biodiversity in the area can be affected. Animal and plant life has been greatly threatened due to flooding in some habitats and water diversion in others; fragmentation of habitat may lead to heavy losses of biological diversity. The river system downstream has also been affected: there has also been a 50% loss in sediment and nutrients downstream, which will cause erosion to river systems, wetlands, and seacoast ecosystems, leading to adversely impacted fisheries and wetland watersheds.

Social impact: About 1.3-1.9 million people were forced to move from their homes along the Yangtze River due to the construction of the Three Gorges Dam, with peasants accounting for 87% of people who live in the reservoir area, and are mostly uneducated. The way in which the Chinese government compensates people forced to move is called the "lump sum" method: people were given grants of the total net worth of their home and land, according to criteria put in place by the government. This method does not always mean equal and effective resettlement as some people were forced to buy homes at a higher price than the amount of money they were given for compensation. In this relocation project, there was only enough land to give 125,000 farmers or farming families, much of the land less fertile than the land to be inundated. Many farmers switched to planting citrus fruits, or other value added products. Estimates state that 140 cities, about 1,000 villages, two cities, and 100,000 acres of fertile farmland will be inundated by the reservoir: this yields 10% of China's annual grain, 50% of which is rice. The thousands of others who were not able to receive new land as part of their compensation were trained for jobs in cities. Since the number of people being relocated is so high, the odds that the peoples' livelihoods can be re-established is low.

Useful search terms: flood control; flooding caused by dam failure; Three-Gorges dam; hydropower; The Grand Renaissance Dam; Ilisu dam; Narmada Sardar Sarovar Dam

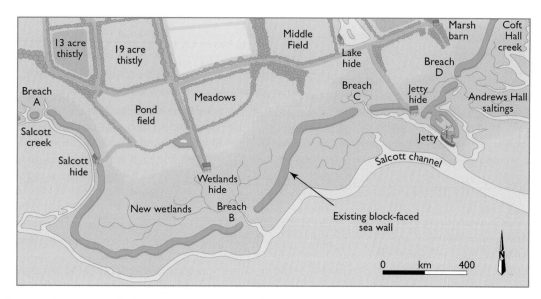

Figure 3.16 Site on the Blackwater estuary in Essex, England, showing a proposed sea wall breach to allow arable land to develop into coastal marshland (*Source:* Harman 2002. Reproduced with permision of ICE Publishing).

Figure 3.17 Natural storm and flood protection provided by mangroves on Iriomote Island, Japan (*Source:* Kzaral on Flickr).

The UK's Met Office and Environment Agency have a devoted Flood Forecasting Centre; combining meteorology and hydrology expertise the Centre provides a specialised hydrometeorology service. The centre forecasts for all natural forms of flooding – river, surface water, tidal/coastal and groundwater.

River level gauges with automated alarms can be a relatively low-cost approach to detecting flood risk through data that is collected from automatic rain gauges, water-level gauges and automatic weather stations. The stations automatically report data to computer base stations. If gauges exceed a predetermined level, emergency management and utility personnel are notified, and wider warnings to the local community can be disseminated to mobile phones through SMS. Fort Collins in Colorado operates a Flood Warning System that is designed to notify the public and emergency services about: real-time information on rainfall, storm water run-off and weather conditions, and early detection of hazardous conditions.

Local knowledge: It is often the case that some of the best information about existential flood risk is overlooked. In rural areas it can be useful to speak to farmers and land managers (but maybe not in the case where these local stakeholders are trying to sell you land to develop upon) and other local residents about historical and recent flood events. Sometimes the names of streets/roads (i.e., Water or Brook Street) and neighbourhoods or new developments (i.e., the Water Meadow Estate or the Marshes) can also be indicative of historic flood risk. Flood markers on bridges and buildings (such as Figure 3.10) also provide useful evidence.

Flood warnings: Flood forecasting is an important component of flood warning, where the distinction between the two is that the outcome of flood forecasting is a set of forecast time-profiles of channel flows or river levels at various locations, while "flood warning" is the task of making use of these forecasts to make decisions about whether warnings of floods should be issued to the general public or whether previous warnings should be rescinded or retracted. See Case Study 3.5 for an example from Nepal.

CASE STUDY 3.5

Early Warning – Saving Lives (Nepal)

The project aimed to establish a community managed Early Warning System (EWS) in flood-prone areas where assets and community resources are lost annually. The EWS consists of five tower systems supporting electrically powered sirens. They run on independent battery-charged power sources that are recharged intermittently from the national grid system, but are not reliant on the grid system. Towers were chosen for the increased distance of view they offered to observers, the increased range they gave to the sirens, and for their psychological impact within the communities. Additionally, a broad range of awareness raising activities, training in EWS use and operation, advice and guidance on long-term sustainability, and the management of the systems took place in the community. The project was able to convince community members of the benefit and value of EWS which was built on their traditional "watch & warn" system for wild life.

    Useful links:

    http://practicalaction.org/dipecho

    http://practicalaction.org/file/region_nepal/early-warning-saving-lives.pdf

Education and awareness raising of flood risk: Public awareness, participation and community support are essential components of good flood risk management. Public participation can not only raise awareness of flood risk; it can also inform decisions and contribute to the successful implementation of actions (such as the need to engage with local communities in risk reduction, emergency management and evacuation procedures). Individuals, businesses and communities can play an important local role in flood management by acting as their own first line of defence against flooding. These actions can play an important role in complementing and supporting the work undertaken by the responsible national and local authorities. It can be particularly beneficial to communities if children (for instance at school) are engaged in these awareness-raising activities as children can be excellent as disseminating the messages at home.

3.13.3 Control of Hazard

Protect natural river levees and/or flood plains: Natural flood protection can be attained by protecting and restoring wetlands and floodplains, and by restoring a river's natural flow and meandering channel. Giving at least some floodplain back to a river will give the river more room to spread out. Furthermore, wetlands act as natural sponges, storing and slowly releasing floodwaters after peak flood flows have passed (see Case Study 3.6). For instance, a single acre of wetland, saturated to a depth of one foot, will retain 330,000 gallons of water enough to flood 13 average-sized homes thigh-deep. Coastal wetlands reduce storm surge and slow its velocity.

CASE STUDY 3.6

Slowing the Flow – A Unique Answer to Flooding in Pickering

As the title suggests, the aim of this pioneering scheme in North Yorkshire, England is to Slow the Flow of flood water through the headwaters to even out damaging flood peaks from a flashy catchment (i.e., floods often rise and then recede very quickly). A range of environmentally friendly and remarkably cost effective flood alleviation solutions are proving useful in the upper catchment of the River Derwent, comprising woody debris dams (see Figure 3.18), mini bunds, riparian tree planting, moorland drain blocking and peat restoration, along with catchment sensitive farming methods. They are combined with a floodwater storage bund above Pickering to provide protection from up to 1:25 year flood events.

Recent government cutbacks have meant that rural areas have largely been precluded from flood defence funding. Thus this is an excellent example of the Environment Agency and other agencies working with the community to provide a locally funded and cost effective scheme that could be used in many other remote flood hit communities, while also reducing the scale (and therefore huge cost) of large downstream urban schemes.

Useful links:

http://www.forestry.gov.uk/fr/infd-7zuclx

https://environmentagency.blog.gov.uk/2015/11/11/slowing-the-flow-working-with-nature-to-reduce-flood-risk-in-north-yorkshire/

Figure 3.18 A woody debris dam used as part of an environmentally friendly and cost effective flood alleviation scheme on the upper catchment of the River Derwent, England (*Source:* Reproduced with the permission of Forestry Commision).

Artificial flood defences/walls/ditches: Traditional structural approaches to managing flood risk have focused upon the construction of different types of large-scale barriers. The Netherlands has arguably been at the forefront of these approaches to the management of flood risk, initially through the use of artificial dikes (dykes, or elevated embankments) and wind powered pumps to protect fields and villages from rivers and coastal areas. As the structures got more extensive and complex councils were formed from people with a common interest in the control of water levels on their land, and so the first water boards began to emerge. Over the last few centuries the Dutch have developed more and more elaborate structural solutions to manage flood risk, including the Zuiderzee Works and the Delta Works. By the end of the twentieth century, all sea inlets had been closed off from the sea by dams and barriers. Central to the Delta Works project was the construction of the Oosterscheldekering, a storm surge barrier across the Oosterschelde estuary which is only closed during storms (see Figure 3.19). It is worth noting that while these approaches to managing flood risk may be effective, they are nonetheless very expensive (in capital and ongoing operation and maintenance costs) and can exacerbate flood risk if they ever fail (due to technological or human error) or indeed if they are not designed to incorporate the impacts of climate change and sea-level rise.

Sustainable urban drainage systems (SUDS): Drainage systems can contribute to sustainable development and improve urban design, by balancing the different issues that influence the development of communities. SUDS mimic nature and typically manage rainfall close to where it falls. SUDS can be designed to slow water down (attenuate) before it enters streams, rivers and other watercourses, they provide areas to store water in natural contours and can be used to allow water to soak (infiltrate) into the ground or evaporated from surface water and lost or transpired from vegetation (known as evapotranspiration) (see Figure 3.20).

Figure 3.19 The Oosterscheldekering surge barrier, Netherlands (*Source:* Vladimír Šiman).

SUDS can use a number of techniques (these are expanded upon in Table 3.7) but it is worth acknowledging that there are some potential drawbacks to the use of SUDS. For instance, the initial costs of SUDS will invariably require some additional capital expenditure as well as longer-term maintenance requirements; the effectiveness, running costs and custodianship (i.e., who is responsible for looking after them) of SUDS is too often unclear. Also, it is possible that SUDS can be viewed as some sort of panacea for dealing with flood risk on new developments but clearly this is problematic as site conditions can limit functionality such as circumstances where there is a high water table level or indeed when developments are located in a riverine flood risk area (Figure 3.21).

3.13.4 Mitigation of Hazard

Localised flood defences/walls/gates:
Town or city level flood defences can be used in conjunction with larger-scale defences to ensure that key strategic locations do not get flooded; sometimes this may result in some developments (i.e., commercial estates and recreational areas) being sacrificed to protect critical locations (such as hospitals and city centres). The Foss Flood Barrier in York is an example of this approach (see Figure 3.22). The Foss Barrier was built in 1987 across the Foss River near its mouth at Castle Mills. When closed, it prevents floodwater from the River Ouse forcing the flow of the Foss back on itself. When the river level reaches 7.8m the barrier is then lowered, after running pumps for several minutes to clear silt and debris from the river bed. It takes four minutes to lower the barrier. To avoid the build-up of water behind the barrier causing the Foss to burst its banks, the water is pumped around

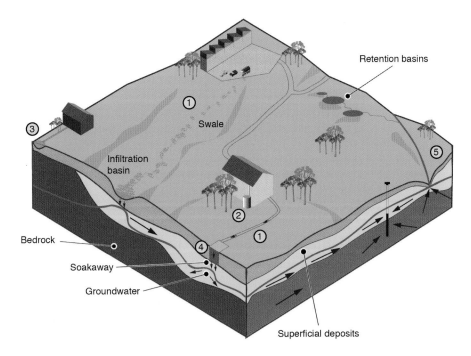

Figure 3.20 Overview of the principles of SUDS. During a storm event, surface water flows through swales and filter trenches that remove pollutants(1). The peak river discharge is delated and reduced by: storage of water for re-use (2), storage in ponds (3), or infiltration of water to the ground through infiltration basins and soakaways (4). This process improves the quality of the water in rivers and decreases peak river discharge (5) (*Source:* WEDC).

the barrier and into the Ouse. However, as with many constructed defences, this approach is not faultless; on 26 December 2015, with the barrier's pumping station flooded, the Environment Agency lifted the barrier – enabling water from the Ouse to flow up into Foss and resulting in more than 600 properties being flooded.

Temporary flood barriers (or 'demountables') are engineered to provide similar levels of protection to permanent flood defences, but with the advantage of being fully removable when not required. Thus, they are useful when flood warnings are feasible and permanent flood walls are undesirable. Case Study 3.7 provides a good example of how useful these types of defences can be.

Flood resilient/resistant materials for buildings: The main principle of flood resilience in this context is to use materials that are compatible with coming into contact with floodwater (see Figure 3.24). There are two main strategies for flood resilience:

- Water exclusion strategy – where emphasis is placed on minimising water entry whilst maintaining structural integrity, and on using materials and construction techniques to facilitate drying and cleaning. This strategy is favoured when low flood water depths are involved (not more than 0.3m). It should be noted that even with this strategy water is still likely to enter the property.
- Water entry strategy – where emphasis is placed on allowing water into the building, facilitating draining and consequent drying. Standard masonry buildings are at significant risk of structural damage if there is a water level difference between outside and inside of about 0.6m or more. This strategy is therefore favoured when potentially high flood water depths are involved (greater than 0.6m).

Table 3.7 SUDS Options Table.

SUDS feature	Description
Green roofs	A planted soil layer is constructed on the roof of a building to create a living surface. Water is stored in the soil layer and absorbed by vegetation.
Rainwater harvesting	Rainwater is collected from the roof of a building or from other paved surfaces and stored in an overground or underground tank for treatment and reuse locally. Water could be used for toilet flushing and irrigation.
Infiltration (percolation) trench	This is typically an excavated long shallow hole filled with gravel. Another similar drainage structure is a French drain, which directs water away from a building foundation, but is usually not designed to protect water quality.
Soakaway	A soakaway is designed to allow water to quickly soak into permeable layers of soil. Constructed like a dry well, an underground pit is dug filled with gravel or rubble. Water can be piped to a soakaway where it will be stored and allowed to gradually seep into the ground. Required area is dependent on runoff volumes and soils
Filter Strip	Filter strips are grassed or planted areas that runoff is allowed to run across to promote infiltration and cleansing. A minimum length of 5 metres is typically required.
Permeable paving	Paving which allows water to soak through. Can be in the form of paving blocks with gaps between solid blocks or porous paving where water filters through the block itself. Water can be stored in the sub-base beneath or allowed to infiltrate into ground below. This paving can typically drain double its area.
Bioretention area	A vegetated area with gravel and sand layers below designed to channel, filter and cleanse water vertically. Water can infiltrate into the ground below or drain to a perforated pipe and be conveyed elsewhere. Bioretention systems can be integrated with tree pits or gardens. Typically the surface area is 5-10% of drained area with storage.
Swale	Swales are vegetated shallow depressions designed to convey and filter water. These can be 'wet' where water gathers above the surface, or 'dry' where water gathers in a gravel layer beneath. Swales can be lined or unlined to allow infiltration. It is necessary to account for width to allow safe maintenance typically 2-3 metres wide.
Hardscape storage	Hardscape water features can be used to store run-off above ground within a constructed container. Storage features can be integrated into public realm areas with a more urban character. These can be located above or below ground and sized to the storage needs.
Pond / Basin	Ponds can be used to store and treat water. 'Wet' ponds have a constant body of water and run-off is additional, while 'dry' ponds are empty during periods without rainfall. Ponds can be designed to allow infiltration into the ground or to store water for a period of time before discharge. The required area is dependent on runoff volumes and soils.
Wetland	Wetlands are shallow vegetated water bodies with a varying water level. Specially selected plant species are used to filter water. Water flows horizontally and is gradually treated before being discharged. Wetlands can be integrated with a natural or hardscape environment. Typically, a 5-15% drainage area is needed to provide good treatment
Underground storage	Water can be stored in tanks, gravel or plastic crates beneath the ground to provide attenuation. The area needed for this type of storage is dependent on runoff volumes and soils.

Flood-resistant defences for buildings and access points: In some cases, it may not be necessary to protect an entire building from flood risk as it may be possible to protect a few potentially weak or vulnerable points. For instance, the access points and stairways to train and metro stations located underground can be particularly vulnerable to the ingress of flood waters. In Tokyo, some of the metro stations have flood barriers located at the top of the access to stairs (see Figure 3.25) that can be swiftly deployed in the event of heavy localised rainfall.

Figure 3.21 SUDS in use at the Building Research Establishment's Innovation Park (Scotland): Examples of swales and permeable paving (*Source: Bosher, 2015*).

Amphibious buildings, or buildings/ roads on stilts: Rising water levels means that the building – or rather, the foundation – floats. Conditions necessary for this type of construction are the presence of surface water and a lightweight building style, for example with a (wooden) skeleton construction method. Floating or amphibian homes must be well anchored into position. This can be done in a number of ways (see Case Study 3.7 for an example). In Maasbommel (the Netherlands) some homes have been built to float up and down along the mooring posts (or what are sometimes referred to as 'dolphins'). In other situations, homes can be fastened to the quay or anchored to the ground with steel cables. Attention must be given to ensuring that no large objects can become lodged under the homes, causing them to tilt when water levels subside. Measures must also be taken against ice build-up.

Figure 3.22 The Foss Flood Barrier in York, shown in the opened position to allow water to pass through (*Source:* Bosher, 2015).

CASE STUDY 3.7

The Demoutable Flood Defences of Bewdley

Bewdley is a town on the River Severn, Britain's longest river, which rises in mid-Wales and flows 350km to its mouth near Bristol. Heavy rainfall in the Welsh mountains and a large catchment lead to regular flooding of the river. In November 2000, Bewdley suffered serious flooding and 140 properties were affected. After the flood, the aim was to construct an effective flood defence for Bewdley, which would also enhance the Georgian town and its eighteenth-century quay. The Environment Agency consulted local people about the best solution to the flooding problem. They decided on an innovative demountable aluminium barrier system called 'the invisible defence' (see Figure 3.23). The barrier works with a below-ground wall, which acts as a cut-off to underground flow during flood events. At a cost of about £7m, 450m of demountable flood defence was built, with 150m of permanent flood defence wall. When a high flow is forecast the Agency Flood Warning gives at least 24 hours to bring in and put up the barriers. It has been reported that it takes 12 Environment Agency staff five hours to erect barriers up to a height of 6m. To date the barriers have been used at least once every year since they were installed.

3.13.5 Emergency Response

Emergency planning: An emergency management plan outlines the actions that should be carried out by different groups or "responders" as an emergency event unfolds. It also provides important background information on the conditions in the affected area. It is important that these plans are tested, regularly updated and shared with the relevant stakeholders. It is increasingly becoming a core responsibility for local government authorities (i.e., in the UK and Australia) to draw up and test

Figure 3.23 Part of the innovative temporary aluminium barrier system being deployed on the banks of the River Severn in Bewdley, England (*Source:* Reproduced with permission of David Throup).

emergency plans, based on regular risk assessments and probabilistic scenarios (including the establishment and dissemination of evacuation plans). From a built environment perspective, it is advisable for developers to liaise with the local emergency planning/management authority when undertaking new developments. At the household and business level, there are a number of emergency preparations that can be undertaken to deal with flood risk (see Table 3.8).

Flood Liaison and Advisory Groups: As an example of good practice, planning policy in Scotland states that each local council should convene a Flood Liaison and Advice Group (FLAG) or combine with other councils to do so, possibly on a catchment basis. The purpose of a FLAG is to act as a forum for the key public and private interests to share knowledge and offer advice on flood risk. FLAGs should be informal advisory groups and those which function well usually meet at regular intervals and include members from several different organisations, each of whom will contribute from their available data, experience and professional judgement.

From within the local authority this is likely to include:

- the planning department
- roads/engineering department,
- emergency planning department
- building control
- possibly landscape/environment department
- the following external agencies will also usually be involved as appropriate:
  - representative of an environment agency
  - representative of a water authority (supply and sewerage)
  - representative from the house building industry
  - representative from the insurance industry, and
  - National Park Authorities, for relevant areas

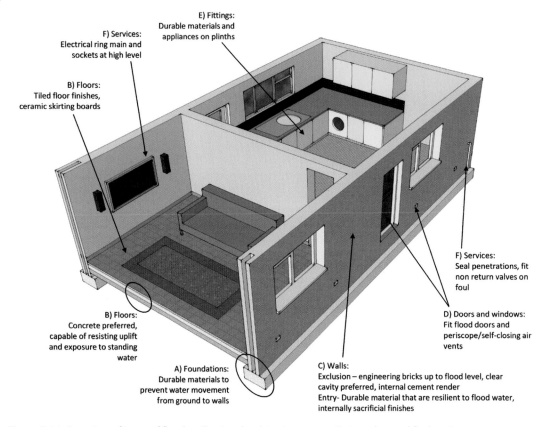

E) Fittings:
Durable materials and
appliances on plinths

F) Services:
Electrical ring main and
sockets at high level

B) Floors:
Tiled floor finishes,
ceramic skirting boards

F) Services:
Seal penetrations, fit
non return valves on
foul

B) Floors:
Concrete preferred,
capable of resisting uplift
and exposure to standing
water

D) Doors and windows:
Fit flood doors and
periscope/self-closing air
vents

A) Foundations:
Durable materials to
prevent water movement
from ground to walls

C) Walls:
Exclusion – engineering bricks up to flood level, clear
cavity preferred, internal cement render
Entry- Durable material that are resilient to flood water,
internally sacrificial finishes

Figure 3.24 Overview of types of flood resilient and resistant measures that can be used for housing (*Source:* Reproduced with permission of Simon Bunn).

Locating critical infrastructure and emergency services/ resources in safe/protected areas sounds like a straightforward common sense option. However, it is not always so clear cut, particularly as much of the developed world's infrastructure has already been built (in some cases, for hundreds of years, when modern-day populations and climates were not considered). This can result in a legacy of poorly planned and conceived infrastructure that (even if not originally) may now be located in hazard prone areas; for instance, due to sea level rise, subsidence of land, denudation of natural barriers, over development, unregulated construction and lack of maintenance of the infrastructure. In these 'legacy' cases, it is important that the vulnerabilities to hazards are identified and suitably assessed, including the interdependencies between different systems that may be owned and managed by different stake-holders (i.e., the dependencies between water treatment and supply and the electricity network).

For new developments it should be possible to meet (or draw them up if not exist) regulations that prohibit new public infrastructure in hazardous areas. In doing so it can be possible to keep utilities, roads, fire stations, police stations, and other infrastructure out of areas that will be affected by flood events and/or will need regular repair. Communities should plan ahead for future infrastructure needs (e.g., schools, police and fire stations), especially in areas of new growth, to ensure that such infrastructure can be sited in appropriate locations.

Figure 3.25 Flood gates at an entrance to the Tokyo Metro system. These barriers can be deployed quickly to protect the Metro system in the event of a potential flood (*Source:* Bosher, 2015).

CASE STUDY 3.8

Amphibious Buildings

Amphibious buildings lie on the ground out of flood periods and are likely to float when the water level rises during flood (see Figure 3.26). They do not therefore float permanently unlike the floating buildings which can be found in many countries in urban areas along lakes or slow-flowing rivers. Examples of these buildings were recently built in Maasbommel (Netherlands). Currently there are 37 two storeys high houses with semi-circular metal roofs and yellow, green or blue facades. The cellar, however, is not built into the earth. The hollow foundation of each house works in the same way as the hull of a ship, buoying the structure up above water. To prevent the floating houses from drifting away, the structures slide up along broad steel posts ('dolphins') – and as the water level sinks, so they sink back down again.

Useful links:

http://www.baca.uk.com/index.php/living-on-water/amphibious-house

http://www.urbangreenbluegrids.com/projects/amphibious-homes-maasbommel-the-netherlands/

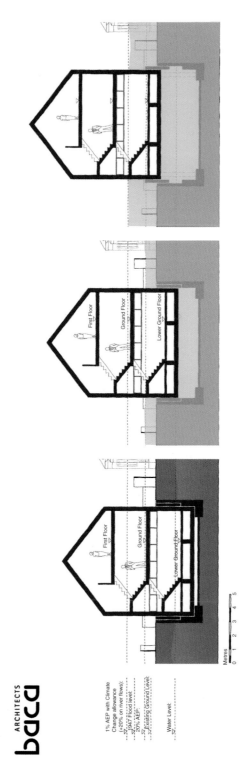

Figure 3.26 Example of an amphibious house, as designed by BACA Architects. When a flood occurs, the entire building rises up in its dock and floats there, buoyed by the floodwater (Source: Reproduced with permission of BACA Architects).

Table 3.8 What to Do and What Not to Do Before, During and After a Flood (adapted and modified from FEMA and EA Flood line).

What to do	What NOT to do
Preparing for a flood	
✓ Check with your local flood control agency to see if your property is at risk from flooding. ✓ Liaise with your local Flood Liaison and Advisory Group (FLAG) if such groups exist. ✓ Purchase flood insurance ✓ Buy sandbags or flood board to block doors ✓ Make up a flood kit including a flash light, blankets, battery-powered radio, first aid kit, rubber gloves, and key personal documents. Keep it upstairs is possible.	✓ Do not underestimate the damage the flood can do.
During the flood (when the flood warning has been issued)	
✓ Listen to the radio or television for information. ✓ Be aware that flash flooding can occur. If there is any possibility of a flash flood, move immediately to higher ground. ✓ Be aware of stream, drainage channels, canyons and other areas known to flood suddenly. Flash floods can occur in these areas with or without typical warnings such as rain clouds or heavy rain. ✓ Secure your home. If you have time, bring in outdoor furniture. Move essential items to an upper floor. ✓ Turn off utilities at the main switches or valves if instructed to do so. Disconnect electrical appliances. ✓ If you have to walk in water, walk where the water is not moving. Use a stick to check the firmness of the ground in front of you.	✓ Do not touch electrical equipment if you are wet or standing in water. ✓ Do not walk through moving water. ✓ Do not drive into flooded areas. If floodwaters rise around your car, abandon the car and move to higher ground if you can do so safely. ✓ Do not camp or park your vehicle along streams, rivers or creeks, particularly during threatening conditions.
After the flood	
✓ Stay away from damaged areas unless your assistance has been specifically requested by police, fire, or relief organization. ✓ Play it safe. Additional flooding or flash floods can occur. Listen for local warnings and information ✓ Return home only when authorities indicate it is safe. ✓ Clean and disinfect everything that got wet.	✓ To not drink or use floodwater; water may be contaminated by oil, gasoline or raw sewage.

3.14 Summary

Floods are naturally occurring processes that can become disasters due to human activities; such as:

- Poor (or unregulated) urban planning (building homes, offices, infrastructure and essential services on flood prone land)
- Buildings and infrastructure not designed to cope with the physical impacts of floods
- Increased urbanisation (leading to more rapid run-off from surfaces to water courses)
- Lack of, or ineffective, urban drainage systems (to cope with expanding developments or impacts of climate change)
- Socio-economic conditions that force poor and marginalised sections of society to live in flood prone areas in cities and villages.

- Poorly conceived river/flood management schemes (leading to over reliance of physical assets that need long-term maintenance and become quickly outdated)
- Deforestation of steep slopes and poor land use practices (leading to instable surfaces and increased run-off)
- The construction of dams (leading to flooded tracts of land, they also pose flood risks due to the potential structural failures of dams)
- Lack of, or ineffective, emergency preparedness procedures.

There is a growing appreciation that floods are largely human-induced and thus there is much that can be done to reduce the impacts of flood and in some circumstances to eliminate flood risk completely.

Key points:
- Most floods are human-made, thus we need to be better at avoiding the creation of flood risk; in simple terms we should not build in flood risk areas unless proportionate flood risk reduction features have been incorporated into the design of the development.
- While invariably it will be impractical, and indeed unnecessary, for all buildings and structures to be built (or retrofitted) to make them flood resilient, it is important for such considerations to be undertaken in the most critical of cases (i.e., related to critical infrastructure, schools and hospitals).
- Holistic multi-hazard multi-stakeholder approaches should be mainstreamed in order to increase the resilience of the built environment. This will invariably include a better appreciation of upstream land use practices and collaboration with the relevant agencies.
- For large-scale developments, it is important to seek the services of a qualified engineer/designer as well as considering the potential benefits of using natural approaches to flood risk management.
- The impacts of climate change should be more scientifically incorporated into the design of flood management infrastructure.

Reinforcement
- What are the differences between pluvial and riverine floods?
- What types of factors can increase the rate of run-off?
- What does a 1:100 flood return period mean?
- Name four different types of features that can be used in a Sustainable Urban Drainage System?

Questions for discussion
- Please name a recent flood that was clearly a 'natural disaster' (i.e., there we no significant human induced influences that contributed towards the disaster)?
- To what extent is climate change linked to increases in flooding events globally?
- In relation to development on flood plains, should the developer be more explicit about telling customers about flood risks or should this responsibility fall on the customer?
- Does affordable flood insurance encourage inappropriate development in flood prone areas?

Further Reading

Books:
Booth, C.A. and Charlesworth, S.M. eds., (2014). *Water Resources in the Built Environment: Management Issues and Solutions*. John Wiley & Sons

Jha, A.K., Bloch, R. and Lamond, J., (2012). *Cities and flooding: a guide to integrated urban flood risk management for the 21st century*. World Bank Publications

Parker, D.J., (2000). *Floods*. Taylor & Francis

Paul, B.K., (2011). *Environmental hazards and disasters: contexts, perspectives and management.* Wiley & Sons

Smith, K., (2013). *Environmental Hazards: Assessing Risk and Reducing Disaster.* (6th Edition), Routledge

White, I., (2013). *Water and the city: Risk, resilience and planning for a sustainable future.* Routledge

Articles and guidance:

Biswas, A.K. (2004) Dams: cornucopia or disaster?, *International Journal of Water Resources Development*, 20:1, 3–14, DOI: 10.1080/0790062032000170571

British Standards Institution BS8582:2013 – Code of Practice for *Surface Water Management for Development Sites*

Building Research Establishment, '*Soakaway Design*' – Digest 365 (BRE 365)

CIRIA C697 The SuDS Manual Woods-Ballard, B.; Kellagher, R. et al

CIRIA R156 Infiltration Drainage – Manual of Good Practice Bettess, R.

ICE (2010) '*Facing up to rising sea levels*', Institution of Civil Engineers/Building Futures, London

Useful websites:

Susdrain www.susdrain.org

CIRIA (UK) www.ciria.org

Building Research Establishment www.bre.co.uk

Department of Homeland Security (US) https://www.ready.gov/floods

FEMA (USA) Flood Mitigation Best Practices Portfolio http://www.fema.gov/mitigation-best-practices-portfolio

4

Windstorms

Views of inundated areas in New Orleans following breaking of the levees surrounding the city as the result of Hurricane Katrina. September 11th, 2005. (*Source:* Lieut. Commander Mark Moran, NOAA Corps, NMAO/AOC).

Disaster Risk Reduction for the Built Environment, First Edition. Lee Bosher and Ksenia Chmutina.
© 2017 John Wiley & Sons Ltd. Published 2017 by John Wiley & Sons Ltd.

Storms have been quoted in many historical texts, particularly in ways that associated such events as forms of punishment from some form of divine entity (such as a god or gods). For instance, in Greek mythology there were several gods of storms: *Aiolos*, keeper of storm-winds, squalls and tempests; *Briareos*, the god of sea storms; and *Aigaios*, a god of the violent sea storms. In reality, windstorms are natural events that are normally benign, but occasionally (some would now say more frequently) they can be so extreme as to result in huge disasters. This chapter will focus on the causes and impacts of the most damaging types of windstorms, namely large low-pressure tropical weather fronts (i.e., hurricanes, cyclones and typhoons) and smaller-scale short-term events such as tornadoes.

4.1   Learning Objectives

By the end of this chapter you will learn:

- What are the different types of windstorms?
- What are the main causes of windstorms?
- What are the typical impacts of windstorms?
- How to identify the risk of windstorms and assess vulnerabilities?
- How to reduce the risk of different types of windstorms using structural and non-structural measures?

4.2   Living with Windstorms

Windstorms can occur pretty much anywhere. Historical records suggest that the most damaging windstorms tend to occur in coastal regions where humankind has historically been attracted to the edge of the sea for access to fishing and trade. In tropical regions of the world, these locations tend to provide environments that are hazardous due to warm sea temperatures (that can be prime breeding grounds for such storms) and low-lying coastal belts that can be vulnerable to the impacts of high winds and associated storm surges (elevated sea levels).

The most disastrous windstorms, most of which have occurred in coastal areas, during the last century are listed in Table 4.1 (deaths) and Table 4.2 (economic damages). These tables provide an interesting overview of the different impacts that similar events can have in different parts of the world. The historical events indicate that the highest amount of deaths have all occurred in Asia (particularly Bangladesh and China) while all of the most economically damaging windstorms (Hurricanes) have affected the USA (during the last 25 years).

Windstorms are created when a centre of low atmospheric pressure develops with a system of high pressure surrounding it. This combination of opposing forces can create winds and result in the formation of storm clouds. There are many varieties and names for storms that are typically based on the scale (i.e., tornado/cyclone), geographical location (i.e., typhoon/hurricane) and underlying climatic conditions (snowstorm/dust storm). The correct classification of these different types of windstorms is important because the risk management approaches required for dealing with different types of storms will invariably need to be different. Thus, it is important to never confuse a tornado with a hurricane or other type of windstorm because tornadoes and hurricanes are different phenomena. Perhaps the only similarity between tornadoes and hurricanes is that they both contain strong rotating winds that can cause damage.

Table 4.1 The 10 Most Deadly Windstorms in the Last Century.

Year	Event	Location	Death toll
1931	Great Bhola Cyclone	Bangladesh	Approx. 500,000
1938	Super Typhoon Nina	China	171,000
1975	Cyclone 02B	Bangladesh	138,866
1935	Cyclone Nargis	Myanmar	138,366
1971	Swatlow	China	100,000
1949	Bangladesh	Bangladesh	61,000
1954	Wenchou	China	50,000
1974	Bangladesh	Bangladesh	36,000
1999	Chittagong	Bangladesh	22,000
1939	Urir	Bangladesh	15,000

Table 4.2 The 10 Most Damaging (in Economic Terms) Windstorms in the Last Century.

Year	Hurricane	Location	Total damages (Billion US$)
2005	Katrina	USA	125
2012	Sandy	USA	50
2008	Ike	USA	30
1992	Andrew	USA	26.5
2004	Ivan	USA	18
2004	Charley	USA	16
2005	Rita	USA	16
2005	Wilma	USA	14.3
2011	Irene	USA	14
2004	Frances	USA	11

4.3 Overview of the Typical Impacts of Windstorms

EM-DAT statistics (Table 4.3) show that in the 50 years between 1964 and 2013 there were 3,300 major windstorms globally, that killed nearly 884,000 people, affected nearly 1 billion people and caused an estimated US$975bn in damages. Typically, for each event, windstorms are the second most prevalent cause of disasters (after floods), they tend to cause the highest amount of economic damages overall and per event are marginally less destructive that earthquakes.

The primary effects of windstorms include loss of life, damage to buildings and other structures, including bridges, electricity transmission systems and roadways.

Secondary effects can be similar across the full range of natural hazards; typically including economic hardship due to a temporary (but sometimes prolonged) decline in business transactions, reduced revenues from tourism and rebuilding costs. The impact on those affected may cause psychological

Table 4.3 Windstorm Averages Per Year Between 1964 and 2013.

Major storms	66
Deaths	17,691
Total people affected	19,107,888
Total economic damages (US$)	19.5bn

damage, in particular where deaths, serious injuries and loss of property occur. In coastal areas, the associated extensive flooding can lead to chronically wet houses, which are linked to an increase in respiratory problems and other illnesses. Coastal flooding, associated with the impacts of storm surges, can also result in the salinisation of the ground and longer-term infertility of agricultural land.

4.4 Causes of Windstorms

Disasters caused by windstorms are often referred to as 'natural disasters', implying that the main causes of the disaster are the natural hazards. In many cases, this is indeed a reasonable comment, with factors such as strength of wind speeds, intensity of rainfall, size of the weather system, topography and local vegetation (to name just a few) playing important roles. However, there is increasing acknowledgement that the impacts of windstorms can be exacerbated by human activities; such as:

- Buildings and infrastructure not designed to cope with the physical impacts of windstorms
- Poorly conceived coastal/flood management schemes (leading to over-reliance of physical assets that need long-term maintenance and become quickly outdated)
- Long-term draining of waterlogged areas that then become protected from the deposition of alluvial soils resulting in land that is subsiding (see Case Study 4.1 for an example of this).
- Deforestation of natural coastal vegetation and forests that can act as natural buffers providing some protection from the impacts of strong winds and storm surges
- Poor (or unregulated) urban planning (building homes, offices, infrastructure and essential services on flood prone land)
- Lack of, or ineffective, emergency preparedness procedures.

All of these factors are human-made and play important roles in how natural processes can turn into hazards that result in disasters. Importantly, these are the key factors that improved by those

---

CASE STUDY 4.1

New Orleans: The Sinking City

Early in the twentieth century, mechanical pumping technology enabled the draining and subdivision of swamps in the growing settlement of New Orleans. However, the reclamation of these waterlogged areas had an unexpected consequence: it made ground levels fall.

This process called subsidence occurred through a number of different mechanisms. Organic matter in the soil oxidised, so soil volume was reduced. As pumping extracted water from the ground, soil particles collapsed onto each other. The removal of the cypress swamps brought an end to soil creation through organic decomposition. Finally, the levees that had been constructed along the length of the Mississippi to stop flooding prevented the replenishment of soil by alluvial material.

By the turn of the twenty-first century, the city had become a giant bowl. Ground levels had fallen to as low as 12 feet (3.6 meters) below sea level; the city was completely surrounded by levees; and the only way to remove water from drains and sewers was by pumping it over the levee to Lake Pontchartrain.

> **What Does the Future Hold?**
>
> The city's surface has subsided, but not at equal rates: some soil types are more susceptible than others. Beyond that, different types of structures respond differently to subsidence. For instance, slab on grade foundations tip and tilt as the ground sinks, but pile foundations remain in place as the earth around them disappears.
>
> The low ground in New Orleans continues to subside, and even the high ground that constitutes approximately half the city must be drained mechanically. These circumstances created havoc during and after Hurricanes Katrina and Rita, when pumps and levees failed. A new approach to water management, one that recognises and adapts to the dynamic ecology of New Orleans' location between the river and the lake, in the world's fourth-largest delta is needed.

involved in how the built environment is planned, designed, built, managed and upgraded by playing a more positive role in addressing disaster risk as an everyday part of their jobs.

4.5 Tropical Windstorms

> **Tropical windstorm** – a rapidly rotating storm system characterised by a low-pressure centre, strong winds and a spiral arrangement of thunderstorms that produce heavy rain. Depending on its location and strength, a tropical cyclone is referred to by names such as hurricane (Atlantic Ocean and Caribbean), typhoon (West Pacific and South China Seas), or cyclone (Indian Ocean and Bay of Bengal). See Figure 4.1 for the global distribution of tropical windstorms.

Key factors:

Main causes: Tropical windstorms typically form over large bodies of relatively warm water. They derive their energy through the evaporation of water from the ocean surface, which ultimately re-condenses into clouds and rain when moist air rises and cools to saturation. As air flows radially inward, it begins to rotate cyclonically (counter-clockwise in the Northern Hemisphere, and clockwise in the Southern Hemisphere).

Characteristics: The rotating winds of a tropical cyclone are a result of the conservation of angular momentum imparted by the Earth's rotation as air flows inwards toward the axis of rotation. As a result, these types of storms rarely form within 5° of the equator. Tropical windstorms are typically between 62 and 1,243 miles (100 and 2,000 km) in diameter. At the inner radius, air begins to ascend to the top of the troposphere and thus a very low-pressure system is generated at the centre of the storm (see Figure 4.2). This low pressure can generate a bulge in the seas surface of up to 10 meters (33 feet) in height (see Figure 4.3). The highest estimated storm surge of 13 meters (43 feet) was generated by Cyclone Mahina in Australia in 1899. Extremely strong winds can be generated up to 157 mph (252 km/h) or even higher; refer to Table 4.4 that summarises the Saffir–Simpson scale for tropical windstorms.

Key impacts: Tropical windstorms generate very high winds that can destroy housing and infrastructure, particularly in low- and middle-income countries where the standard of construction may be of basic standard or non-engineered. Intense widespread rainfall and associated storm surges from the sea can result in floods that take days (or even weeks) to subside, but effective warnings can help to minimise the loss of lives. Buildings and critical infrastructure can be damaged or destroyed, and it can become extremely difficult for affected areas to function for a prolonged period; economic losses can be vast (see Case Study 4.2). Chemicals and other hazardous substances (such as sewage) can contaminate water bodies and people's homes.

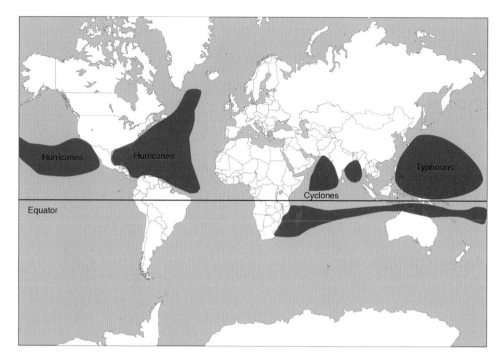

Figure 4.1 Global distribution of tropical windstorms (*Source:* WEDC).

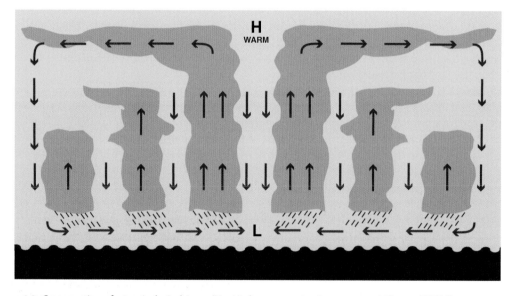

Figure 4.2 Cross section of a tropical windstorm (H = High pressure; L = Low pressure) (*Source:* WEDC).

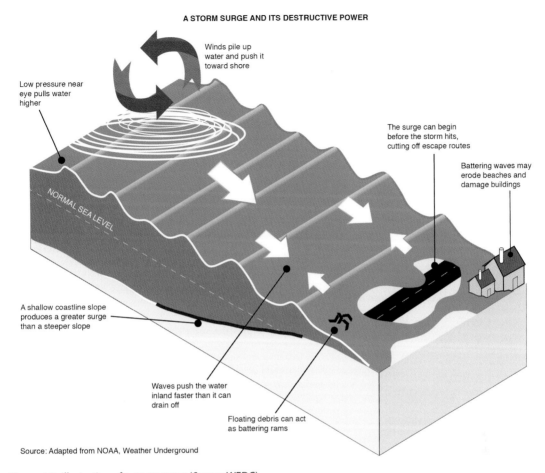

A STORM SURGE AND ITS DESTRUCTIVE POWER

Low pressure near eye pulls water higher

Winds pile up water and push it toward shore

The surge can begin before the storm hits, cutting off escape routes

Battering waves may erode beaches and damage buildings

NORMAL SEA LEVEL

A shallow coastline slope produces a greater surge than a steeper slope

Waves push the water inland faster than it can drain off

Floating debris can act as battering rams

Source: Adapted from NOAA, Weather Underground

Figure 4.3 Illustration of a storm surge (*Source:* WEDC).

Location: We know where the most severe windstorms can strike as these locations tend to have been affected in the (recent) past. Understanding the historical instances of tropical windstorms can give a rudimentary indication of what coastal areas are prone to the impacts of windstorms; particularly in high-risk areas such as low-lying coastal belts, river deltas and wetlands. However, it is worth noting that the frequency of severe tropical windstorms appears to be on the increase and in some cases reaching into higher latitudes so historical data should not be overly relied upon.

Advance warnings: One of the positive attributes of tropical windstorms is that the use of modern metrological technologies makes it possible to identify where tropical windstorms have been generated over the ocean and then to track them to ascertain whether they may pose a risk to inhabited areas. While the internal wind speeds of tropical windstorms can be very high, they do, however, move relatively slowly across the surface of the ocean, thus it is getting increasingly viable to predict where they may make landfall. One or two days advance warning can enable local authorities to warn residents to protect their property or indeed initiate mass evacuation procedures (see Case Study 4.3 for a positive example of effective evacuation mechanisms).

Table 4.4 Saffir–Simpson Hurricane Wind Scale.

Category	Sustained Winds	Types of Damage Due to Hurricane Winds
5 (major)	157 mph or higher 137 knots or higher 252 km/h or higher	**Catastrophic damage will occur:** A high percentage of framed homes will be destroyed, with total roof failure and wall collapse. Fallen trees and power poles will isolate residential areas. Power outages will last for weeks to possibly months. Most of the area will be uninhabitable for weeks or months.
4 (major)	130–156 mph 113–136 knots 209–251 km/h	**Catastrophic damage will occur:** Well-built framed homes can sustain severe damage with loss of most of the roof structure and/or some exterior walls. Most trees will be snapped or uprooted and power poles downed. Fallen trees and power poles will isolate residential areas. Power outages will last weeks to possibly months. Most of the area will be uninhabitable for weeks or months.
3 (major)	111–129 mph 96–112 knots 178–208 km/h	**Devastating damage will occur:** Well-built framed homes may incur major damage or removal of roof decking and gable ends. Many trees will be snapped or uprooted, blocking numerous roads. Electricity and water will be unavailable for several days to weeks after the storm passes.
2	96–110 mph 83–95 knots 154–177 km/h	**Extremely dangerous winds will cause extensive damage:** Well-constructed frame homes could sustain major roof and siding damage. Many shallowly rooted trees will be snapped or uprooted and block numerous roads. Near-total power loss is expected with outages that could last from several days to weeks.
1	74–95 mph 64–82 knots 119–153 km/h	**Very dangerous winds will produce some damage:** Well-constructed frame homes could have damage to roof, shingles, vinyl siding and gutters. Large branches of trees will snap and shallowly rooted trees may be toppled. Extensive damage to power lines and poles likely will result in power outages that could last a few to several days.

CASE STUDY 4.2

Hurricane (Superstorm) Sandy

New York City, lying in the zone where cold Canadian Arctic air masses meet the warm Gulf Stream current, is prone to the effects of tropical storms and hurricanes. 2012's Hurricane Sandy did not break wind or rainfall records in New York, yet it was one of the costliest ever to affect the USA. The surge flooded large parts of lower Manhattan, including subways and tunnels, caused mass power outages and destroyed thousands of homes and businesses in New York and neighbouring New Jersey.

Sandy left western Africa on 11 October 2012 as a cluster of thunderstorms and moved westward to the Caribbean Sea. It then gradually intensified to become a tropical storm and moved towards Jamaica where on 24 October it officially became a hurricane. Sandy then crossed eastern Cuba moving towards the Bahamas and made landfall near Atlantic City, New Jersey on 29 October with sustained surface winds at landfall close to 80 mph and gusts between 85 and 95 mph. Sandy's arrival into the U.S. coast also coincided with both high tide and spring tide. In New York City, this added an extra 20 to 50 cm to the high water mark.

At least 286 people were killed either directly or indirectly by Sandy. There were 147 direct deaths: 72 in the USA and the rest mainly in Caribbean, including 54 in Haiti and 11 in Cuba. In the USA of the 87 indirect deaths from Sandy, at least 50 were attributable to either falls by the elderly, carbon monoxide poisoning from inadequately ventilated generators and cooking equipment, or hypothermia as a cold snap followed Sandy and extended power outages left people without heating. Damage estimated to be over US$71 billion. In New York City, economic losses are estimated at exceeding US$18 billion. Overall, 24 states were affected by Sandy.

In many instances, New York's hazard mitigation measures proved to be lacking, particularly around critical energy infrastructure, where flood walls designed to prevent the ingress of water proved inadequate and ineffective. Much of this damage was predicted by a 2011 report led by Columbia University which considered the impact of climate change upon the city. In particular, the vulnerability of large areas of the city to modest rises in sea level, as well as the potential for transport systems to flood.

The New York example illustrates a failure to adequately assess the vulnerability of the city to the flooding/storm surge hazard in many incidents, but more importantly to act upon earlier vulnerability assessments. It is apparent that as result of its location New York has a fundamental site vulnerability that makes it particularly susceptible to this sort of event. However, the problem was magnified by design weaknesses including the sitting of critical infrastructure in particularly vulnerable locations and unchecked development in vulnerable locations. Whilst in many instances New York did prove resilient and in many areas was able to carry on business as usual, had vulnerabilities in transport and power systems been acting upon, the impact would have been much less.

Search terms: Hurricane Sandy; Superstorm Sandy; New York resilience;

CASE STUDY 4.3

Cyclone Phailin – How Evacuations Can Save Lives

Cyclone Phailin, one of the of the most severe cyclones to have ever affected the North Indian Ocean and Bay of Bengal made landfall on 11 November 2013 near Gopalpur in the state of Odisha (formerly Orissa).

Officials from Odisha's state government said that around 12 million people could be affected. As part of the preparations, 600 buildings were identified as cyclone shelters and people were evacuated from areas near the coast. The cyclone prompted one of India's biggest evacuations with more than 550,000 people moved from the coastline in Odisha and Andhra Pradesh to safer places.

In Odisha, the government issued a high alert to the districts that could potentially be affected, and cancelled the holidays of employees of all 30 districts of the state, asking them to ensure the safety of people. Food and relief materials were stocked up at storm shelters across the state. Support from defence personnel, particularly the Air Force and Navy, for rescue and relief operations was also requested in advance from the National Government. Odisha government had made arrangements for over 1,000,560 food packets for relief. Indian Air Force helicopters were kept on standby in West Bengal to move in for help at short notice.

The evacuation was acknowledged by the UNISDR as a 'landmark success' action in limiting the number of deaths directly caused by the cyclone to 21, compared to over 10,000 deaths when a similar scale cyclone affected the same area in 1999. The loss of lives in Phailin was minimised by effective early warning communication supported by joint efforts from the community, volunteer organisations, local and national levels of government and by donors, but also by the level of preparedness the community and local and national government exhibited.

Search terms: Cyclone Phailin; DRR and evacuation in India; Odisha preparedness

4.6   Tornadoes

> **Tornado** is a violently rotating column of air (also may be referred to as a 'twister') that is in contact with both the surface of the earth and a cumulonimbus cloud or, in rare cases, the base of a cumulus cloud.

Key factors:

Main causes: Tornadoes are considered nature's most violent storms. Generated by powerful thunderstorms, tornadoes can cause fatalities and devastate a neighbourhood in seconds. Tornadoes have been observed on every continent except Antarctica. However, the vast majority of tornadoes occur in what is often refereed as 'Tornado Alley' in the USA, although they can occur nearly anywhere across North America. Tornadoes also occasionally occur in south-central and eastern Asia, northern and east-central South America, Southern Africa, north-western and south-eastern Europe, western and south-eastern Australia, and New Zealand (see Figure 4.4).

Characteristics: Tornadoes are smaller in diameter but move across the ground much faster than tropical cyclones/hurricanes. A tornado appears as a rotating, funnel-shaped cloud that extends from a thunderstorm to the ground with whirling winds that can reach 300 miles per hour (see Figure 4.5). Damage paths can be in excess of one mile wide and 50 miles long. Most tornadoes have wind speeds less than 110 miles per hour (180 km/h), are about 250 feet (80 m) across, and travel a few miles (several kilometres) before dissipating. The most extreme tornadoes can attain wind speeds of more than 300 miles per hour (480 km/h), stretch more than two miles (3 km) across, and stay on the ground for dozens of miles. There are several scales for rating the strength of tornadoes. The Fujita scale rates tornadoes by damage caused and has been replaced in some countries by the updated Enhanced Fujita Scale (refer to Table 4.5).

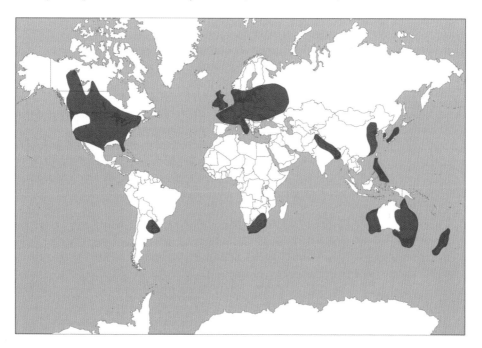

Figure 4.4 Tornado hazard map (*Source:* WEDC).

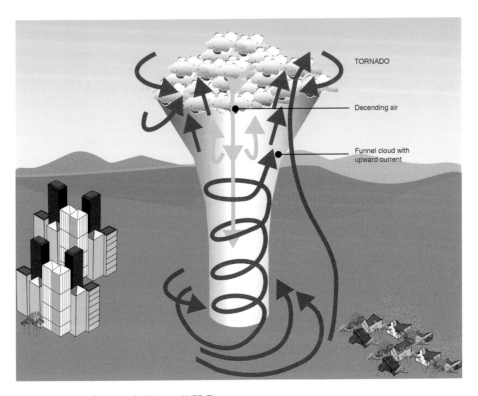

Figure 4.5 Cross section of a tornado (*Source:* WEDC).

Key impacts: Two tornadoes that look almost exactly the same can produce drastically different effects. Also, two tornadoes which look very different can produce similar damage. This is due to the fact that tornadoes form by several different mechanisms, and also that they follow a life cycle which causes the same tornado to change in appearance over time. Typically, an EF3 (significant) scale tornado can pose a serious risk to life and limb. Very few components of an affected buildings are left standing; well-built structures lose all outer and some inner walls. Unanchored homes are swept away, and homes with poor anchoring may collapse entirely. Small vehicles and similarly sized objects are lifted off the ground and transported in the air as dangerous projectiles. Wooded areas will suffer almost total loss of vegetation, and some tree debarking may occur. Statistically speaking, EF3 is the maximum level that allows for reasonably effective residential sheltering in place in a first floor interior room, closest to the centre of the house (the most widespread tornado sheltering procedure in America for those with no basement). At the most extreme end of the damage scale, the impacts of an EF5 tornado represents the upper limit of tornado power, and destruction is almost always total. An EF5 tornado can rip well-built, well-anchored homes off their foundations and into the air. Even large, steel reinforced structures can be completely levelled. Despite their relative rarity, the damage caused by EF5 tornadoes represents a disproportionately extreme hazard to life and limb—since 1950 in the United States, only 58 tornadoes (0.1% of all reports) have been designated EF5, and yet these have been responsible for more than 1,300 deaths and 14,000 injuries (21.5% and 13.6%, respectively).

Table 4.5 Enhanced Fujita Scale.

Scale	Wind speed (estimated)			Relative frequency	Potential damage
	mph	km/h	m/s		
EF5	>200	>322	>90	<0.1%	Total destruction of buildings. Strong framed, well-built houses levelled off foundations and swept away; steel-reinforced concrete structures are critically damaged; tall buildings collapse or have severe structural deformations; some cars, trucks and trains can be thrown approximately 1 mile (1.6 kilometres).
EF4	166–200	267–322	74–90	0.70%	Extreme damage. Well-constructed and whole frame houses completely levelled; cars and other large objects thrown and small missiles generated.
EF3	136–165	218–266	61–73	3.40%	Entire stories of well-constructed houses destroyed; severe damage to large buildings such as shopping malls; trains overturned; trees debarked; heavy cars lifted off the ground and thrown; structures with weak foundations are badly damaged.
EF2	111–135	178–217	50–60	10.70%	Considerable damage. Roofs torn off well-constructed houses; foundations of frame homes shifted; mobile homes completely destroyed; large trees snapped or uprooted; light-object missiles generated; cars lifted off ground.
EF1	86–110	138–177	38–49	31.60%	Moderate damage. Roofs severely stripped; mobile homes overturned or badly damaged; loss of exterior doors; windows and other glass broken.
EF0	65–85	105–137	29–37	53.50%	Minor or no damage. Peels surface off some roofs; some damage to gutters or siding; branches broken off trees; shallow-rooted trees pushed over. Confirmed tornadoes with no reported damage (i.e., those that remain in open fields) are always rated EF0.

Location: The exact location of tornadoes can be hard to predict, particularly in a timely manner that can enable local populations to evacuate or seek shelter. We can identify the general geographical regions where tornadoes can strike as they tend to have been affected in the (recent) past (see Figure 4.4). Larger windstorms such as hurricanes typically show up on satellites days beforehand, but the conditions that favour tornados appear much more quickly and unexpectedly. One of the most notorious regions for tornado activity is 'Tornado Alley' that extends from northern Texas, Oklahoma, Kansas, into Nebraska and increasingly likely to include the Canadian prairies (see Case Study 4.4 for an example of a tornado that occurred outside of 'Tornado Alley').

Advance warnings: The paths of tornadoes are smaller and they last for shorter periods of time, so predicting a particular tornado requires a fine-grain understanding that is more difficult for scientists. Meteorologists can identify conditions that are likely to lead to severe storms. They can issue warnings when atmospheric conditions are right for the development of tornadoes. They can use radar to track the path of thunderstorms that might produce tornadoes. However, it is still not possible to detect a funnel cloud by radar and predict its path, touchdown point, and other important details. Much progress has recently been made in the detection of tornadoes using Doppler radar. It is important to bear in mind that even a 10-minute warning can be sufficient for a family/community to seek shelter but this is only really a viable option if there are protected safe zones in buildings or cellars where people can seek refuge in.

Doppler radar is a specialised radar that uses the Doppler effect to produce velocity data about distant objects. It does this by bouncing a microwave signal off a desired target and analysing how the object's motion has altered the frequency of the returned signal.

CASE STUDY 4.4

March 2012 Tornados in the USA

The March 2–3, 2012 tornado outbreak was one of the deadliest outbreaks recorded so early in the year. It occurred over a large section of the Southern United States into the Ohio Valley region and resulted in 41 tornado-related fatalities, 22 of which occurred in Kentucky. During these two days 279 tornado warnings were issued.

The outbreak began early in the morning on the 2nd of March, with an initial round of storms and thunder associated with the incoming warm front attached to a rapidly deepening low-pressure area over the central Great Lakes. The initial round of storms allowed for a strong warm air mass to enter the region, with temperatures rising to near-record levels for early March and instability combining with extreme wind shear, resulting in a highly volatile air mass. As a result, a second, much larger broken line of discrete supercells developed and followed the Ohio River, with additional storms developing farther south.

The deadliest tornado of the outbreak was an EF4 that carved a 79 km path of damage from Fredericksburg, Indiana to Bedford, Kentucky. Along its track, the tornado destroyed hundreds of homes and killed 11 people. The tornado destroyed many homes and businesses, with some of the buildings being levelled and swept away. Transportation was thrown through the air, hundreds of trees were snapped and debarked along the damage path.

Search terms: tornado; supercells

4.7   Risk Management

4.7.1   Historical Approaches

Generally, people have been quite effective at adapting to the hazards posed by windstorms, mainly because windstorm prone areas were good for fishing (i.e., coastal regions) and agricultural use (such as tornado-prone plains). Some of these adaptations have included:

- Using flexible timber-framed buildings designed with large windows or even no/minimal sides so that the air pressure differences between in the internal and external environment are more quickly equalised (see Figure 4.6 for an example of a Polynesian *fale*).
- Protecting thatched roofing with old fishing nets that are anchored into the ground (see Figure 4.7).
- Building houses elevated above the usual ground level to minimise likelihood of inundation from coastal storm surges and pluvial rain (see Figure 4.8), and
- Development of indigenous early warning systems (i.e., in regions of India, Bangladesh and Burma surrounding the Bay of Bengal) that use the unusual behaviour of the wind, tidal reaches, and even animals as precursors to impending storms.

Figure 4.6 Typical faleo'o beach fale, Manono Island, Samoa (Source: Reproduced with permission of Sonja Pieper).

Figure 4.7 Traditional thatched 'kutcha' hut with fishing net used as storm protection. A practice used in the cyclone prone East Godavari region of Andhra Pradesh, India (*Source:* Bosher, 2002).

However, over the last century cities have flourished in economic importance, and population growth has resulted in situations where development has been unregulated, poorly planned and inadequately built with little regard for traditional approaches to predicting (see Thinking Point 4.1) and dealing with the risk of windstorms.

Figure 4.8 Traditional high status house in coastal area of Sri Potti Sri Ramulu, Nellore District of Andhra Pradesh, India. The house is built on an elevated plinth to protect the property from localised flooding caused by storm surges (*Source:* Bosher, 2002).

Thinking Point 4.1

History of Weather Forecasting

The tradition of weather forecasting dates back to the earliest civilisations that used reoccurring astronomical and meteorological events as a basis for monitoring seasonal changes in weather. Around 650 BC, the Babylonians tried to predict short-term weather changes based on the appearance of clouds and optical phenomena such as haloes. By 300 BC, Chinese astronomers had developed a calendar that divided the year into 24 festivals, each festival associated with a different type of weather event. *Meteorologica*, a philosophical treatise that included theories about the formation of rain, clouds, hail, wind, thunder, lightning, and hurricanes, was written by Greek philosopher Aristotle around 340 BC. Some of his observations were remarkably accurate, although there were also some significant errors. This text was a basis for weather forecasting until seventeenth century.

From the fifteenth century onwards it became clear that personal observations were not enough for predicting the weather, and that instrumentation that could measure the properties of the atmosphere, such as moisture, temperature, and pressure, were needed. The first known design for a hygrometer was described by Nicholas Cusa in the mid-fifteenth century. Galileo invented an early thermometer in 1592, and Evangelista Torricelli invented the barometer for measuring atmospheric pressure in 1643.

The invention of the telegraph and the emergence of telegraph networks in the mid-nineteenth century allowed the routine transmission of weather observations to and from observers and compilers. This led to the creation of weather maps, so that surface wind patterns and storm systems could be identified and studied. Weather-observing stations began appearing all across the globe, eventually spawning the birth of synoptic weather forecasting, based on the compilation and analysis of many observations taken simultaneously over a wide area, in the 1860s.

With the formation of regional and global meteorological observation networks in the nineteenth and twentieth centuries, more data were becoming available for observation-based weather forecasting. A great stride in monitoring weather at high altitudes was made in the 1920s with the invention of the radiosonde. Small lightweight boxes equipped with weather instruments and a radio transmitter, radiosondes are carried high into the atmosphere by a hydrogen or helium-filled balloon that ascends to an altitude of about 30 km before bursting. During the ascent, these instruments transmit temperature, moisture, and pressure data (called soundings) back to a ground station. There, the data are processed and made available for constructing weather maps or insertion into computer models for weather prediction. Today, radiosondes are launched every 12 hours from hundreds of ground stations all over the world.

4.8   Hazard Identification

There are a number of ways to identify the hazard of windstorms. The most straightforward method is to identify the historical prevalence of hurricanes, cyclone and tornadoes and understand the impacts of such events for different locations (see Figure 4.9 for a global overview for between 1945 and 2006). Section 4.1 of this chapter highlights that typically windstorms occur in certain parts of the world and that the geographical location of these events has a bearing on whether large windstorms are called hurricanes (Atlantic Ocean and Caribbean), typhoons (West Pacific and South China Seas), or cyclones (Indian Ocean and Bay of Bengal). However, over-reliance upon historical events can be problematic as there can be a complex number of factors that influence the generation and magnitude of large windstorms; for instance, the impacts of climate change and cyclical climatic

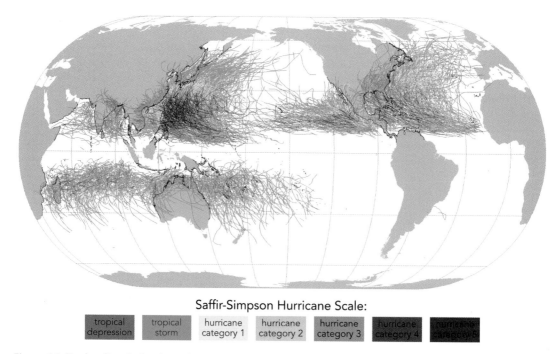

Saffir-Simpson Hurricane Scale:

| tropical depression | tropical storm | hurricane category 1 | hurricane category 2 | hurricane category 3 | hurricane category 4 | hurricane category 5 |

Figure 4.9 Tracks of tropical cyclones, hurricanes and typhoons between 1945 and 2006 (*Source:* Citynoise).

events such as El Niño and La Nina. For instance, the National Oceanic and Atmospheric Administration (NOAA) reported that 2015 experienced a below-normal Atlantic hurricane season while on the other hand the eastern and central Pacific seasons shattered records with an above-normal amount of storms and hurricanes (see Case Study 4.4).

The National Oceanic and Atmospheric Administration (NOAA) is an American scientific agency within the United States Department of Commerce that looks at the conditions of the oceans and the atmosphere. NOAA has been at the forefront globally in developing systems for warning of dangerous weather as well as in developing guides to improve understanding and stewardship of the environment. NOAA's National Hurricane Centre (NHC) uses a range of analysis tools (see Table 4.6) for identifying problematic weather fronts, for predicting the generation of strong storms and for monitoring the paths of such storms.

There are a number of scientific and non-scientific ways in which windstorm related hazards can be identified, tracked and assessed. Some of the most prominent methods are summarised in Table 4.6.

Table 4.6 Overview of Typical Hazard Identification Methods for Windstorms.

Context	Hazard identification methods
Historical experience	We know where major windstorms can strike because certain locations have been affected in the (recent) past. These incidents might have been recorded and reported by national or local institutions, such as NOAA in the USA, the Met Office in the UK and the Meteorological Department in India, which have compiled detailed records of previous windstorms and also have the capacity to monitor real time storm activity (see below).
Meteorological mapping and monitoring	National and global atmospheric and meteorological observation and monitoring is now very sophisticated. On a day-to-day basis this body of science can inform us of what the weather forecast will be for a particular region or city but in more recent years this information has contributed towards the prediction of (some) impending weather problems. For instance, the identification low-pressure weather fronts over warm tropical seas can provide the breeding ground for large windstorms; as the pressure continues to drop the system can be monitored to ascertain whether a storm is being generated (for information about how these storms are named please see 'Thinking Point 4.1'). NOAA has a range of tools that can be used independently or collectively to identify and monitor windstorms, such as (for more information search for 'NOAA analysis tools'): • Hovmöller Diagram (5-Day Satellite), • Upper-Air Time Sections, • GFS Pressure Change Analysis, • ASCAT Ocean Wind Data, • Streamlines, Sea Surface Temperature
Predication and warning	Scientific advances in the last few decades have contributed towards the development of accurate storm prediction capabilities. Low-pressure weather fronts can be tracked and computer models can be used to predict the possible path and location of potential landfall of storm (see Figure 4.10 for an example of the predicted path of Hurricane Katrina in 2005). These predictions can be more accurate and timely for large storms such as hurricanes and cyclones and less predictable for tornados. Accurate predications can be vital in enabling national and local governmental authorities to issue evacuation procedures and to put in place (previously developed and tested) emergency plans. This information can also be useful for generating 'what-if' scenarios as part of emergency planning testing and in identifying the most vulnerable areas, people or infrastructure. Figure 4.11 provides an example of a searchable hurricane risk map of New York, indicating high-risk areas and locations of nearest evacuation centres.

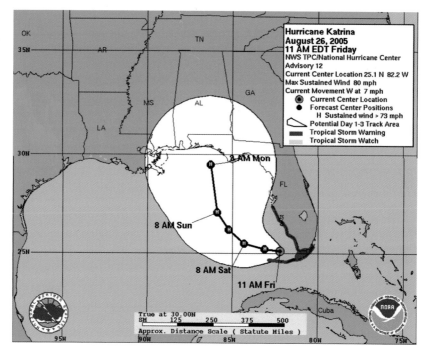

Figure 4.10 Predicted path of Hurricane Katrina in 2005 that turned out to be quite accurate (*Source:* NOAA).

4.9   Assessment of the Vulnerabilities

Once an understanding has been obtained of what the hazards might be, it is then critical to gain an appreciation of what types of sites, buildings, facilities and services could be affected by the identified hazards. This involves an assessment of the potential vulnerabilities of specific aspects of such locations that can be categorised as follows (and previously detailed in Chapter 2):

1. Physical vulnerability
2. Social vulnerability
3. Economic vulnerability
4. Environmental vulnerability
5. Governance vulnerability

1. **Physical Vulnerability** may be determined by aspects such as population density levels, the location and gradient of the site, design and materials used for critical infrastructure and for housing.
   Example: Wooden homes are less likely to collapse in an earthquake, but they can be more vulnerable to the impacts of severe windstorms, particularly tornadoes.
2. **Social Vulnerability** refers to the root causes that make it harder for some people (i.e., women, the elderly, ethnic groups) to withstand adverse impacts of storm related hazards.
   Example: Some people have limited choices about where to live, especially if they have migrated to large cities, and may end up inhabiting hazard prone locations (i.e., low-lying coastal areas); thus when storms occur some citizens, particularly children, women, elderly and the less mobile,

CASE STUDY 4.5

The 2015 Windstorm Season and the Impacts of El Niño

The Atlantic hurricane season produced 11 named storms, including four hurricanes (Danny, Fred, Joaquin and Kate), two of which, Danny and Joaquin, became major hurricanes. Although no hurricanes made landfall in the United States this year, two tropical storms – Ana and Bill – struck the north-eastern coast of South Carolina and Texas, respectively. Ana caused minor wind damage, beach erosion and one direct death in North Carolina, and Bill produced heavy rain and flooding while it moved across eastern Texas and Oklahoma.

NOAA scientists credit El Niño as the leading climate factor influencing both the Atlantic and Pacific seasons this year. "El Niño produces a see-saw effect, suppressing the Atlantic season while strengthening the eastern and central Pacific hurricane seasons," said Dr Gerry Bell, a lead seasonal hurricane forecaster at NOAA's Climate Prediction Centre. "El Niño intensified into a strong event during the summer and significantly impacted all three hurricanes seasons during their peak months."

Bell said El Niño suppressed the Atlantic season by producing strong vertical wind shear combined with increased atmospheric stability, stronger sinking motion and drier air across the tropical Atlantic, all of which make it difficult for tropical storms and hurricanes to form and strengthen. However, El Niño fuelled the eastern and central Pacific seasons this year with the weakest vertical wind shear on record.

The eastern Pacific saw 18 named storms, including 13 hurricanes, nine of which became major. 2015 was the first year since reliable record keeping began in 1971 that the eastern Pacific saw nine major hurricanes. Hurricane Patricia was the strongest hurricane on record in the Western Hemisphere in terms of maximum wind speed at 200 miles per hour (322 km/h) and lowest air pressure at 879 millibars. Hurricane Sandra, which formed at the tail end of the season, was the strongest hurricane in the eastern Pacific so late in the year, with a maximum sustained wind speed of 145 miles per hour (233 km/h).

The central Pacific shattered its records too, with 14 named storms, including eight hurricanes, five of which became major hurricanes, the most active season since reliable record-keeping began in 1971. Three major hurricanes (Ignacio, Kilo and Jimena) churned at the same time east of the International Dateline, the first time that was ever recorded.

Source: NOAA, 2015

See Thinking Point 4.1 for an explanation of why and how storms are named.

Useful search terms: NOAA, super storms, El Niño, Hurricane Katrina, Hurricane Andrew, Typhoon Phailin

may be unable to protect themselves or evacuate if necessary. In relation to Hurricane Katrina in 2005, some of the most affected areas of New Orleans were the poorest; for instance, the Lower Ninth Ward, a predominately African-American neighbourhood.

3. **Economic Vulnerability.** Levels of vulnerability can be highly dependent upon the economic status of individuals, communities and nations.

Example: Some families may live hand-to-mouth existences and struggle to save funds that could help them in times of crisis (such as floods and storms but also more common/everyday crises such as ill health and unemployment). The inability of some of the poorer sections of New Orleans' society to evacuate due to a lack of transportation played a key factor in why they were so disproportionately affected by Hurricane Katrina.

Thinking Point 4.2

How to Name a Cyclone?

The reason for giving storms and tropical cyclones names is practical: it is much easier to remember a name than a number or a technical term, and thus avoid confusion when passing the information between hundreds of widely scattered stations, coastal bases, and ships at sea. In order to have an organised and efficient naming system, it was decided to identify storms using names in alphabetical order: thus a storm with a name, which begins with A, like Anne, would be the first storm to occur in the year. Since 1953, Atlantic tropical storms have been named from lists originated by the National Hurricane Centre, which is now maintained and updated by an international committee of the World Meteorological Organisation. The original name lists featured only women's names, but in 1979 men's names were introduced and they now alternate with the women's names. Six lists are used in rotation (thus, the list used in 2014 will be used again in 2020); these lists differ depending on the origin of the storm (Caribbean Sea, Gulf of Mexico and the North Atlantic; Eastern North Pacific; Central North Pacific; and Western North Pacific and the South China Sea Names). The names are used one after the other. When the bottom of one list is reached, the next name is the top of the next list.

Here's an example of the list used in the Caribbean Sea, Gulf of Mexico and the North Atlantic, Eastern North Pacific:

2016	2017	2018	2019	2020
Alex	Arlene	Alberto	Andrea	Arthur
Bonnie	Bret	Beryl	Barry	Bertha
Colin	Cindy	Chris	Chantal	Cristobal
Danielle	Don	Debby	Dorian	Dolly
Earl	Emily	Ernesto	Erin	Edouard
Fiona	Franklin	Florence	Fernand	Fay
Gaston	Gert	Gordon	Gabrielle	Gonzalo
Hermine	Harvey	Helene	Humberto	Hanna
Ian	Irma	Isaac	Imelda	Isaias
Julia	Jose	Joyce	Jerry	Josephine
Karl	Katia	Kirk	Karen	Kyle
Lisa	Lee	Leslie	Lorenzo	Laura
Matthew	Maria	Michael	Melissa	Marco
Nicole	Nate	Nadine	Nestor	Nana
Otto	Ophelia	Oscar	Olga	Omar
Paula	Philippe	Patty	Pablo	Paulette
Richard	Rina	Rafael	Rebekah	Rene
Shary	Sean	Sara	Sebastien	Sally
Tobias	Tammy	Tony	Tanya	Teddy
Virginie	Vince	Valerie	Van	Vicky
Walter	Whitney	William	Wendy	Wilfred

The only time that there is a change in the list is if a storm is so deadly or costly that the future use of its name on a different storm would be inappropriate for reasons of sensitivity. Storms names, such as Katrina (USA, 2005), Mitch (Honduras, 1998) and Tracy (Darwin, 1974), are examples of this.

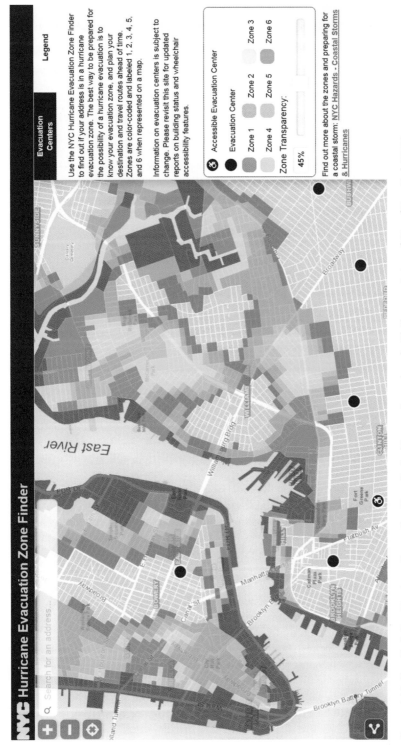

Figure 4.11 New York City: Hurricane Evacuation Zone finder (Source: Reproduced with the permission of City of New York 2016).

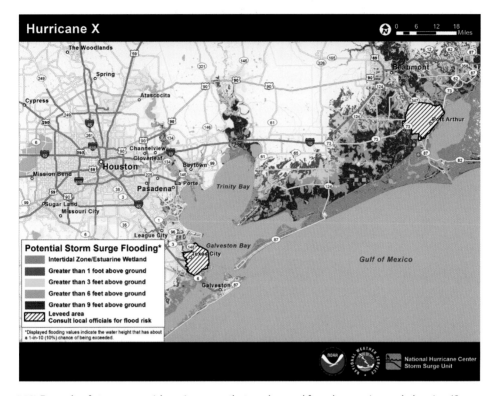

Figure 4.12 Example of storm surge risk zoning maps that can be used for urban zoning and planning (*Source:* NOAA, 2015).

4. **Environmental Vulnerability.** Natural resource depletion and resource degradation are key aspects of environmental vulnerability.

   Example: Mangrove swamps along the coast of India have historically afforded some protection for coastal areas from storms and coastal floods. However, in recent years many mangrove swamps have become denuded due to locals using the mangrove trees for firewood (see Figure 3.11) but also on a larger scale, coastal mangroves have been destroyed to made 'aquaculture farms' where shrimps/prawns can be bred in large artificially constructed tanks containing saline water.

5. **Governance Vulnerability.** The complex roles of national and local government in producing or addressing disaster risk can too often become overlooked.

   Example: Note the contrast between two neighbouring counties – India and Burma – and their capabilities to cope with similar-scale tropical cyclones. Cyclone Phailin in 2013 was one of the largest cyclones in the region and prompted one of India's biggest evacuations with more than 550,000 people in Odisha and Andhra Pradesh moved to safer places inland from the coastline. The Government was acclaimed because the evacuation was acknowledged as a 'landmark success' action in limiting the number of deaths directly caused by the cyclone to 21, compared to over 10,000 deaths when a similar-scale cyclone affected the same area in 1999. In contrast, the Burmese Government's lack of preparedness and support for humanitarian aid in the aftermath of Cyclone Nargis in 2008 turned a disaster into a broader public health crisis resulting in 138,366 deaths.

4.10   Determination of the Risk

Figure 4.12 illustrates how storm surge risk mapping can be used to encourage the siting of sensitive or critical services (i.e., houses and emergency services) in low-risk areas and also limit developments in high-risk to less-sensitive recreational activities.

Effects associated with windstorms can be divided into primary and secondary. The primary effects of windstorms are those caused by severely high wind speeds, the debris that the winds can carry and actual contact with flood water, resulting in death due to damage to buildings and infrastructure, drowning, and so on. The secondary effects are longer-term effects that are indirectly related to the impacts of the flooding caused by the heavy rain and the storm surges from the sea (see Table 4.7 for more examples). Scientists have developed ways to calculate risks, on the basis of, for example, wind speed calculations, surge height statistics, strength of current defences, estimates of the amount of affected population and economic loss estimates.

4.10.1   Windstorm Damage Estimation

Windstorm damage estimation can be a very useful exercise in the management of storm risk by informing the decision making process. Damage estimation is also used extensively in the insurance industry, in order to calculate the financial exposure an organisation or development may have to a windstorm event (as well as associated events such as flooding). For instance, the Europe Windstorm Models (version 15) and North Atlantic Hurricane Models developed by Risk Management Solutions provide capabilities to manage coastal storm risk with the latest science and data on hurricane activity. The models also provide the ability to explore how climate variability impacts the view of

Table 4.7 Examples of the Primary and Secondary Effects of Windstorms.

	Primary effects	Secondary effects
Physical	*Damage to...* • Critical infrastructure • Houses • Offices/shops • Schools • Vehicles • Equipment • Agriculture	*Resulting in....* • Homelessness/displacement • Destroyed/disrupted essential services • Sedimentation of water channels/sewers • Disrupted education provision • Polluted land (due to salinization of soil) • Loss of crops
Wellbeing/ health	*Deaths or injuries to...* • People • Livestock	*Leading to...* • Epidemic diseases (i.e., cholera) • Post-traumatic stress disorder • Unhygienic conditions • Polluted land/homes • Strains upon social support services
Economic	*Immediate...* • Repair/replacement costs • Loss of business	*Longer term...* • Rehabilitation costs • Insurance costs (and increased premiums) • Reduced commerce • Loss of tourist revenues • Reduced land values

Table 4.8 Summary of the Viability of Risk Reduction Options for Addressing Windstorms.

Type of risk reduction option for hurricanes/cyclones/typhoons					
Inherent safety	Prevention	Detection	Control	Mitigation	Emergency response
N	N	Y	#	Y	Y

Type of risk reduction option for tornadoes					
Inherent safety	Prevention	Detection	Control	Mitigation	Emergency response
N	N	N	N	Y	#

Please note:
'Y' – indicates that there are possibly a range of useful risk options available
'#' – indicates that some risk reduction options can be used but they are likely to be of only limited effectiveness
'N' – indicates that other than relocating the built asset there is little that can be done to reduce this hazard/threat

risk so that more informed decisions can be made about the viability of proposed developments and the risk management of new or existing developments.

4.11 Identification and Prioritisation of Risk Reduction Options

As discussed in previous chapters, there are a number of ways to reduce the risk of a hazard. In the case of windstorms, various options are available and these will now be discussed.

Fundamentally, when it comes to dealing with windstorms such as typhoons, cyclones and hurricanes, the risk reduction options will need to consider the impacts of strong winds, storm surges and heavy localised rain. When dealing with the more geographically localised tornadoes the risk reduction options will need to consider the impacts of extremely strong winds and the impacts of hailstones.

Ideally, risk reduction measures should adopt the five interrelated stages as shown in Table 4.8, which summarises the extent to which each of the risk reduction options can be utilised for different types of windstorms. From this summary, it appears that little can be done to detect and control the impacts of tornadoes. A more detailed list of specific risk reduction examples is provided for dealing with windstorm risks in Table 4.9.

4.11.1 Detection of Hazard

Cyclones/hurricanes/typhoons can be detected and tracked days in advance of them making landfall and causing damage. This can provide essential intelligence that can contribute toward accurate forecasting and evacuation planning.

Tornadoes are much harder to accurately predict. However, meteorologists can issue warnings when atmospheric conditions are right for the development of tornadoes. Much progress has recently been made in the detection of tornadoes using Doppler radar. It is important to bear in mind that even a 10-minute warning can be sufficient for a family/community to seek shelter but this is only really a viable option if there are protected safe zones in buildings or cellars where people can seek refuge in.

Table 4.9 Indicative Examples of Risk Reduction Options for Addressing Windstorms.

Risk reduction option	Examples
1) Inherent safety – eliminate the possibility of hazards occurring	It is not possible to avoid development in areas where hurricanes or tornadoes may occur
2) Prevention – reduce the likelihood of hazards	It is not possible to reduce the likelihood of hurricanes or tornadoes occurring
3) Detection – measures for early warning of hazards	*Cyclones/hurricanes/typhoons* can be detected and tracked days in advance of them making landfall and causing damage. This can provide essential intelligence that can contribute toward accurate forecasting and evacuation planning. *Tornadoes* are much harder to accurately predict. However, meteorologists can issue warnings when atmospheric conditions are right for the development of tornadoes. Much progress has recently been made in the detection of tornadoes using Doppler radar. It is important to bear in mind that even a 10-minute warning can be sufficient for a family/community to seek shelter but this is only really a viable option if there are protected safe zones in buildings or cellars where people can seek refuge in.
4) Control – limiting the size of the hazards	*Cyclones/hurricanes/typhoons* Protect natural coastal vegetation such as mangrove forests that can help to take the energy out of the storms and act as a buffer for coastal flooding. Land use zoning could be considered for low lying coastal areas where storm surges are possible (i.e., restricting these areas to non-essential or recreational developments). *Tornadoes* cannot be controlled
5) Mitigation and adaptation – protection from the effects of hazards	*Cyclones/hurricanes/typhoons* Localised storm surge defences/walls/gates/shutters Windstorm resilience/resistant materials for buildings and infrastructure Flood resistant defences for buildings and infrastructure Land use zoning (avoiding exposed/flood prone land). Developing an outreach program about windstorm and mitigation activities in homes, schools, and businesses Non-return valves on sewer pipes *Tornadoes* Localised windstorm proof shutters (for windows and doors) Construction of safe room(s) in homes/offices/factories/schools for 'invacuation' purposes Construction of reinforced concrete basements/bunkers for shelter Tornado proof buildings and infrastructure Land use zoning (avoiding exposed/flood prone land). Developing an outreach program about tornado impacts and evacuation activities for homes, schools, and businesses
6) Emergency response – planning for evacuation emergency access	Land use zoning (i.e., locating essential lifelines services in non-vulnerable locations) Use of realistic 'worse case' scenarios for testing evacuation procedures Emergency planning that is updated and regularly tested through exercises Evacuation route planning and free mass evacuation transportation provision Locating critical infrastructure and emergency services/resources in safe/protected areas

4.11.2   Control of Hazard

There is little evidence to suggest that tornadoes can be controlled. However, for cyclones/hurricanes/typhoons it is argued that the impacts of these storms have become more severe in recent years due to the impact of deforestation in some coastal areas. Therefore the protection, or indeed replanting, of natural coastal vegetation such as mangrove forests should be encouraged (see Figure 4.13 for different examples of how vegetation and constructed defences could be utilised). These natural buffers can help to take the energy out of storms and act as a buffer for coastal flooding. Land use zoning could be considered for low-lying coastal areas where storm surges are possible (i.e., restricting these areas to non-essential or recreational developments).

4.11.3   Mitigation of the Hazard

4.11.3.1   Cyclones/Hurricanes/Typhoons

- Localised storm surge defences/walls/gates/shutters
- Windstorm resilience/resistant materials for buildings and infrastructure
- Flood resistant defences for buildings and infrastructure (see examples in Chapter 3)
- Land use zoning (avoiding exposed/flood prone land).
- Developing an outreach program about windstorm and mitigation activities in homes, schools, and businesses
- Non-return valves on sewer pipes

4.11.3.2   Tornadoes (many of these options will also be useful for mitigating other windstorm hazards)

- Localised windstorm proof shutters (for windows and doors)
- Construction of safe room(s) in homes/offices/factories/schools for 'invacuation' purposes (see Case Study 4.5)
- Construction of reinforced concrete basements/bunkers for shelter
- Tornado-proof buildings and infrastructure
- Land use zoning (avoiding exposed/flood prone land).
- Developing an outreach program about tornado impacts and evacuation activities for homes, schools and businesses

4.11.4   Emergency Response

- Land use zoning (i.e., locating essential lifelines services in non-vulnerable locations)
- Use of realistic 'worse case' scenarios for testing evacuation procedures (Table 4.10)
- Emergency planning that is updated and regularly tested through exercises. This could include undertaking assessments of which sections of society may be more vulnerable and/or exposed to the hazard (see Figure 4.16).
- Evacuation route planning and free mass evacuation transportation provision
- Locating critical infrastructure and emergency services/resources in safe/protected areas

Minimal defence
Many communities have developed right along the ocean with only minimal natural defences from a small strip of beach between them and the ocean.

Natural
Natural habitats that can provide storm and coastal flooding protection include salt marsh, oyster and coral reefs, mangroves, seagrasses, dunes, and barrier islands. A combination of natural habitats can be used to provide more protection, as seen in this figure. Communities could restore or create a barrier island, followed by oyster reefs and salt marsh. Temporary infrastructure (such as a removable sea wall) can protect natural infrastructure as it gets established.

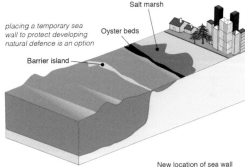

Managed realignment
Natural infrastructure can be used to protect built infrastructure in order to help the built infrastructure have a longer lifetime and to provide more storm protection benefits. In managed realignment, communities are moving sea walls farther away from the ocean edge, closer to the community and allowing natural infrastructure to recruit between the ocean edge and the sea wall.

Hybrid
In the hybrid approach, specific built infrastructure, such as removable sea walls or openable flood gates (as shown here) are installed simultaneously with restored or created natural infrastructure, such as salt marsh and oyster reefs. Other options include moving houses away from the water and/or raising them on stilts. The natural infrastructure provides key storm protection benefits for small to medium storms and then when a large storm is expected, the built infrastructure is used for additional protection.

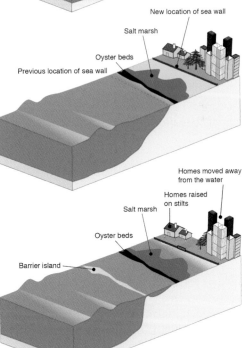

Approaches to coast defence

Figure 4.13 Examples of natural, constructed and hybrid approaches to coastal defence (*Source:* WEDC at Loughborough University adapted from NOAA, 2015).

CASE STUDY 4.6

Building 'Safe Rooms'

A residential safe room is a small, specially designed ('hardened') room, such as a bathroom or closet, or other space within the house that is intended to provide a place of refuge only for the people who live in the house. In areas subject to extreme-wind events, homeowners should consider building a residential safe room.

Having a safe room in a home or small business can help provide protection for the family from injury or death caused by the dangerous forces of extreme winds (see Figure 4.14). Based on current knowledge of tornadoes and hurricanes, the occupants of a safe room built according to the guidance provided by organisations such as FEMA will have a high probability of being protected from injury or death. Current knowledge of tornadoes and hurricanes is based on numerous meteorological records as well as extensive investigations of damage to structures from extreme winds. Having a safe room in each home can also relieve some of the anxiety created by the threat of an oncoming tornado or hurricane (see FEMA P-320 for relevant guidance and Figure 4.15 for an example design plan). Safe rooms in small businesses (or in residences with greater than 16 occupants) are considered community safe rooms and, therefore, must be designed with additional architectural, fire safety, ventilation and other requirements, (as described in FEMA P-361).

Useful FEMA links:

FEMA P-320, *Taking Shelter from the Storm: Building a Safe Room for Your Home or Small Business*

FEMA P-361, *Safe Rooms for Tornadoes and Hurricanes: Guidance for Community and Residential Safe Rooms*

For more information on residential safe room doors and anchoring requirements, the following documents can be downloaded from the FEMA website:

- Residential Tornado Safe Room Doors Fact Sheet.
- Foundations and Anchoring Criteria for Safe Rooms Fact Sheet

---

4.12 Summary

4.12.1 Summary

Severe windstorms are natural processes that are impossible to prevent and extremely difficult to mitigate for. These hazards can become particularly disastrous events due to human activities; such as:

- Buildings and infrastructure not designed to cope with the physical impacts of windstorms
- Poorly conceived coastal/flood management schemes (leading to over reliance of physical assets that need long-term maintenance and become quickly outdated)
- Long-term draining of waterlogged areas that then become protected from the deposition of alluvial soils resulting in land that is subsiding.
- Deforestation of natural coastal vegetation and forests that can act as natural buffers providing some protection from the impacts of strong winds and storm surges
- Poor (or unregulated) urban planning (building homes, offices, infrastructure and essential services on flood prone land)
- Lack of, or ineffective, emergency preparedness procedures.

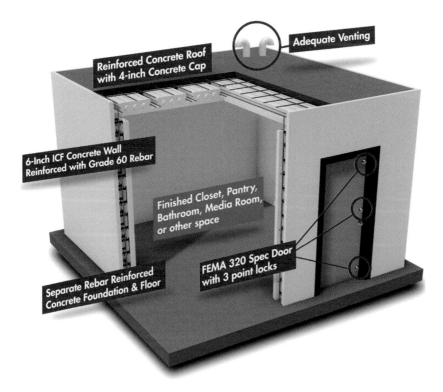

Figure 4.14 Example of residential safe room (*Source:* BuildBlock Building Systems).

Even though we cannot prevent these hazards from occurring, there is still actually quite a lot that we (as construction practitioners, home owners and individuals) can do to reduce the impacts of windstorms.

Key points:

- While invariably it will be impractical, and indeed unnecessary, for all buildings and structures to be built (or retrofitted) to stringent codes, it is important for such considerations to be undertaken in the most critical of cases (i.e., related to key infrastructure, schools and hospitals).
- Holistic multi-hazard multi-stakeholder approaches should be mainstreamed in order to increase the resilience of the built environment
- Awareness of the structural and non-structural approaches required to prepare for, and reduce the impacts of, windstorms needs to be improved across construction disciplines as well as within the general public.
- If in doubt, seek the services of a qualified engineer/designer and contact your local emergency planning authority.

Reinforcement

- What is the 'scale' called that hurricane intensity is measured on?
- What is the 'scale' called that tornado intensity is measured on?
- Why are tornadoes so difficult to predict (in both their location and when they may occur)?
- Why can storm surges be so damaging?

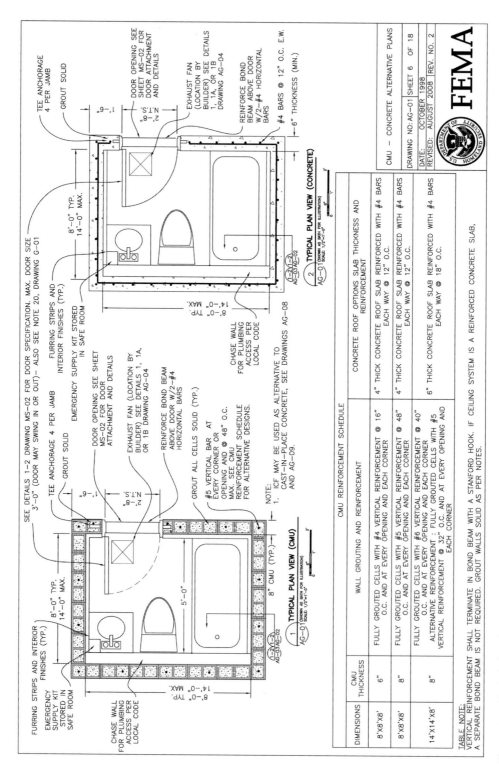

Figure 4.15 Typical plans for a concrete masonry unit (CMU) safe room (FEMA, 1998).

Table 4.10 What To Do and What Not To Do Before, During and After a Windstorm (adapted and modified from FEMA guidance).

What to do	What NOT to do
Preparing for a windstorm	
• Check with the local authority agency to see if the property development is at risk from windstorms and associated flooding. If in doubt, ask neighbours/local residents • Purchase relevant insurance. • Invest in windstorm shutters (accredited ones tested to relevant specifications), alternatively it may be possible to use pressure or impact rated windows. • Invest in household level flood protection • Make up an emergency kit including a flash light, blankets, battery-powered radio, first aid kit, rubber gloves, and key personal documents and medication. Keep this in a safe room/location.	• Do not underestimate the damage the windstorm can do. • Do not assume a property/development is covered for windstorms in a standard insurance policy; check what event are and are not covered for
During the windstorm (when and after a windstorm warning has been issued)	
• Listen to the radio or television for information. • Secure your home. If you have time, bring in outdoor furniture. Move essential items to an upper floor. • Turn off utilities at the main switches or valves if instructed to do so. Disconnect electrical appliances. • If you have to walk in water, walk where the water is not moving. Use a stick to check the firmness of the ground in front of you.	• Do not ignore evacuation warnings unless you have a safe room to take shelter in. • Do not wait outside for the storm to reach your property (i.e., to film it or just out of curiosity) • Do not touch electrical equipment if you are wet or standing in water. • Do not walk or drive through moving water. • Do not camp or park your vehicle along streams, rivers or creeks, particularly during threatening conditions.
After the windstorm	
• Stay away from damaged areas unless your assistance has been specifically requested by police, fire, or relief organization. • Play it safe. Listen for local warnings and information to check that the windstorm has passed. • If you have evacuated, then return home only when authorities indicate it is safe. • Clean and disinfect everything that got flooded.	• To not drink or use floodwater; water may be contaminated by oil, gasoline or raw sewage.

Questions for discussion

- Should extremely low lying areas of cities that are prone to windstorms (such as in New Orleans) be redeveloped or abandoned?
- To what extent are El Niño events linked to increases or decreases in windstorm activity globally?
- What are the best ways of raising awareness of what people should do in the event of a hurricane/typhoon/cyclone?

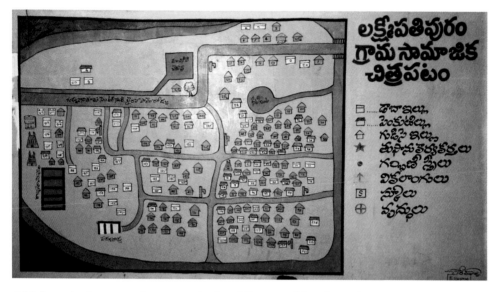

Figure 4.16 Example of a community generated vulnerability map of the village of Laxmipathipuram in Andhra Pradesh, India, indicating households/areas most at risk of the impacts of a cyclone and flood (*Source:* Bosher, 2003).

Further Reading

Books

Doe, R.K., (2015). *Extreme Weather: Forty Years of the Tornado and Storm Research Organisation*. John Wiley & Sons.

Smith, K., (2013). *Environmental Hazards: Assessing Risk and Reducing Disaster* (6th Edition), Routledge

Steinberg, T., (2006). *Acts of God: The Unnatural History of Natural Disaster in America*, Oxford University Press.

Articles/guidance

Elliott, J.R. and Pais, J., (2006). 'Race, class, and Hurricane Katrina: Social differences in human responses to disaster', *Social Science Research*, 35(2), pp. 295–321

EN, BS. (2005). *Eurocode 1: Actions on structures—General actions—Wind actions*

FEMA P-320, xxxx *Taking Shelter from the Storm: Building a Safe Room for Your Home or Small Business*, FEMA

FEMA P-361, xxxx*Safe Rooms for Tornadoes and Hurricanes: Guidance for Community and Residential Safe Rooms*, FEMA

Landsea, C. W., B. A. Harper, K. Hoarau, and J. A. Knaff. (2006). 'Can We Detect Trends in Extreme Tropical Cyclones?, *Science*, 313: 452–454

Mendelsohn, R., Emanuel, K., Chonabayashi, S. and Bakkensen, L., (2012). 'The impact of climate change on global tropical cyclone damage', *Nature Climate Change*, 2(3), pp. 205–209

Olshansky, R.B., (2006). 'Planning after hurricane Katrina', *Journal of the American Planning Association*, 72(2), pp. 147–153

Shanmugasundaram, J., Arunachalam, S., Gomathinayagam, S., Lakshmanan, N. and Harikrishna, P., (2000). 'Cyclone damage to buildings and structures—A case study', *Journal of Wind Engineering and Industrial Aerodynamics*, 84(3), pp. 369–380

Useful websites:

http://www.fema.gov/
http://www.noaa.gov/
http://www.nhc.noaa.gov/
http://www.rms.com
http://www.nhc.noaa.gov/analysis_tools.php

Section III

Geological Hazards

5

Earthquakes

Opening image: A deserted village, abandoned after the 1968 Belice earthquake in Sicily
(*Source:* Lee Bosher, 2014).

Disaster Risk Reduction for the Built Environment, First Edition. Lee Bosher and Ksenia Chmutina.
© 2017 John Wiley & Sons Ltd. Published 2017 by John Wiley & Sons Ltd.

We can do little to reduce the hazard embodied in an active fault or a major earthquake, but we can do a lot about the risk to the structures that we design and build. It is important to remember the frequently quoted observation that earthquakes do not kill, but collapsed buildings and facilities do.

Professor Thomas O'Rourke, Cornell University

Ancient texts, ruined former settlements and damaged historical buildings have provided sufficient evidence that earthquakes have blighted the development of some human habitations for thousands of years. More recent data suggests that earthquakes typically account for the highest proportion of deaths per year than any other types of geophysical or hydrometeorological hazards. Some of these deaths are caused directly by the earthquake but others are caused by the knock-on impacts of the earthquake, such as tsunamis (discussed in this chapter) and landslides (to be discussed in Chapter 7).

LEARNING OBJECTIVES

By the end of this chapter you will learn:

- What are the main causes of earthquakes?
- What are the typical impacts of earthquakes?
- How to identify seismic risks and assess vulnerabilities?
- How to reduce the impacts of earthquakes using structural and non-structural measures?

5.1 Living with Earthquakes

Historical records have proven testament to the long-term fragile relationship between the establishment of human settlements in earthquake prone locations and the impacts of seismic activity. For instance, highly developed ancient centres of scholarship and industry, such as Troy, Mycenae and Knossos, are reported to have been severely affected (if not actually destroyed) by earthquakes over 3,000 years ago. In more recent years (Table 5.1), high profile earthquakes have devastated regions of Haiti, China, Pakistan and Indonesia, and while the most prominent impacts reported in the media tend to be the physical impacts, the socio-political and economic impacts also play an important role in how nations and communities cope with, and recover from, such events.

There is, however, interesting evidence suggesting that historically people living in seismically active regions have found ways to adapt to the threat of earthquakes and some of these approaches will be discussed in this chapter.

5.1.1 Overview of the Typical Impacts of Earthquakes

According to records compiled by EM-DAT (Table 5.2), in the 50 years between 1964 and 2013 there were 998 major earthquakes globally, that killed a total of over 1 million people, affected 174 million people and caused an estimated US$534bn in damages.

The primary effects of earthquakes include loss of life, damage to buildings and other structures, including bridges, power and water distribution systems, roadways, and canals.

Secondary effects include economic hardship due to a temporary decline in business transactions, reduced revenues from tourism and rebuilding costs. The impact on those affected may cause psychological damage, in particular where deaths, serious injuries and loss of property occur; these fears can be exacerbated in areas where earthquake 'aftershocks' are experienced. This can also lead

Table 5.1 The 10 Most Deadly Earthquakes in the Last 50 Years.

Country	Magnitude	Year	Total deaths
China	7.5	1976	242,000
Haiti	7.0	2010	222,570
Indonesia	9.1	2004	165,708
China	7.9	2008	87,476
Pakistan	7.6	2005	73,338
Peru	7.9	1970	66,794
Iran	7.4	1990	40,000
Sri Lanka	9.1	2004	35,399
Iran	6.6	2003	26,796
Iran	7.8	1978	25,000

Note: some deaths may have been caused by tsunamis and landslides generated by the earthquake

Table 5.2 Earthquake Averages Per Year Between 1964 and 2013.

Earthquake events	20
Deaths	21,490
Total people affected	3,487,384
Total economic damages (US$)	10.7bn

to the evacuation of long-established habitations with residents (sometimes forcibly) being relocated elsewhere (see Case Study 5.1). Severe earthquakes can lead to a varying degree of damage, from structures that are entirely destroyed to others that may look safe and functional but in fact may not be structurally sound. Fires can spread due to gas pipe explosions and landslides may also occur. **Tsunamis** may cause extensive flooding in coastal areas.

5.2 Causes of Earthquakes

Disasters caused by earthquakes are often referred to as 'natural disasters', implying that the main cause of the disaster is the natural hazard. In many cases, this is indeed a reasonable assertion, with factors such as geology and topography playing important roles. However, there is increasing acknowledgement that earthquakes are natural processes that become disastrous events due to human activities, such as:

- Poor (or unregulated) urban planning (building homes, offices, infrastructure and essential services in seismically active locations) (see Case Study 5.2)
- Buildings and infrastructure not designed to cope with the physical impacts of earthquakes
- Artificial development of land converted from wetland areas or the sea; such land can be prone to the impacts of liquefaction (see Case Study 5.3)
- Lack of, or ineffective, emergency preparedness and evacuation procedures.

CASE STUDY 5.1

The Abandonment of a Sicilian Town

On the night of 15th January 1968, a large earthquake raked through the Valle del Belice in southwest Sicily. Around 900 people died and 10 towns and villages were significantly damaged. Occupying a panoramic position overlooking the gently undulating valley below, Poggioreale was destroyed when the earthquake struck (see Figure 5.1). After assessing the damage, it was decided that rebuilding the town on the existing site would be 'uneconomical' and so a new area was proposed a few kilometres down the valley. Architects soon got to work on the construction of the new Poggioreale. Contemporary town-planning ideas and techniques were employed for the project and as far as the townsfolk were concerned, their new town was at least close to their old homes, which were visible across the valley and thus the Poggiorealesi were not a significant distance from the land many of them worked.

The old Poggioreale was left as it was, frozen in time, a ghost town shrouded with melancholy, staring out across the countryside at its modern (but rather dystopian) successor. Wandering around the streets of old Poggioreale one can see how the site remains important to the Poggiorealesi, with evidence of gardens and memorials being regularly attended to. Indeed a few hours recently spent walking around both the old and new towns of Poggioreale left the author feeling that there was much more love and affection felt for the old town in comparison to the newer, rather brutal modern town.

Figure 5.1 The deserted main square of Poggioreale; a village abandoned after the 1968 Belice earthquake in Sicily (*Source:* Lee Bosher, 2014).

CASE STUDY 5.2

Haiti Earthquake

On the 12th January 2010 Haiti was struck by a magnitude 7 earthquake with an epicentre 16 miles West of Port-au-Prince and a shallow focus of 5 miles. As a result of this earthquake, 316,000 people were killed and 1 million people were made homeless; 250,000 homes and 30,000 other buildings, including 60% of government buildings, were either destroyed or badly damaged (see Figure 5.2); over 50 hospitals and more than 1,300 were badly damaged, as well as transport and communication infrastructure; and 4,000 inmates escaped from the destroyed prison. The large number of deaths meant that hospitals and morgues became full and bodies then had to be piled up on the streets. Damaged infrastructure hampered rescue and subsequent aid efforts.

One of the main reasons for such destruction was the low-quality building stock and lack of enforced building standards. Structures were often built wherever they could fit within an increasingly urbanised landscape; some buildings were built on slopes with insufficient foundations or steel supports.

Rescue efforts started immediately after the earthquake, and were often performed by the survivors. Later, there was a large international humanitarian response with many countries providing aid and sending military personnel. Six months after the quake as much as 98% of the rubble remained uncleared, making most of the main transport routes in the capital impassable. The number of people receiving humanitarian relief since the earthquake was 1.6 million, and almost no transitional housing had been built. Most of the camps had no electricity, running water, or sewage disposal. Crime in the camps was widespread, especially against women and girls.

In October 2013, the Haitian government launched the country's first national housing policy in a bid to address the shortage of 500,000 new homes that it estimated Haiti needs by 2020. There are, however, concerns that the government-led new housing projects will be negatively affected by corruption and mismanagement, as well as by poor coordination and lack of funds. Disputes over land ownership have proved a major obstacle too, as the earthquake destroyed the majority of title deeds and land registry records.

Lessons learnt from this earthquake illustrated the importance and significant benefits of sustained dialogue and interaction between military actors and humanitarian agencies, and emphasises the need to empower poor individuals and communities to help 'build back better' and to review methods of information collection and management, including the use of social networking to facilitate communications.

Search terms: Haiti earthquake; response and recovery in Haiti; DRR and crowd-sourcing; impact on community; failed international response; failed donor aid

All of these factors are human-made and play important roles in how natural processes such as seismic activity can result in disasters. In explicitly acknowledging the human-made factors that contribute towards earthquake-related disasters it will hopefully make it harder for society to merely

Figure 5.2 Extensive damage to housing caused by the Haitian earthquake in 2010 (*Source:* WEDC).

Case Study 5.3

Christchurch Earthquake and the Impacts of Liquefaction

On 22nd February 2011 the Canterbury region of New Zealand's South Island was struck by a magnitude 6.3 earthquake. It caused significant damages to the second most populous New Zealand city of Christchurch that had recently been affected by another earthquake in 2010. 185 people died as a result of the 2011 Christchurch earthquake. Most of the fatalities were from the collapse of two multi-storey office buildings – the Canterbury Television and Pyne Gould Corporation buildings. The other fatalities were caused by falling bricks, rocks and masonry.

The earthquake caused widespread damage across Christchurch, especially in the central city and eastern suburbs, which were also affected by significant liquefaction problems that produced 400,000 tonnes of silt. Shaking turned water-saturated layers of sand and silt beneath the surface into sludge that gushed upwards through cracks. Properties and streets were buried in thick layers of silt, and water and sewage from broken pipes flooded streets. The foundations of many houses cracked and buckled, making it necessary to demolish those houses afterwards. Christchurch's central business district remained cordoned off for more than two years after the earthquake. Electricity was restored to 75% of the city within three days, but water supplies and sewerage systems took several years to restore in some areas affected by liquefaction.

One of the main lessons learnt was that of the city's foundations. Much of Christchurch was once swampland, beach dune sand, estuaries and lagoons that were drained as the area was settled and developed upon. As a result, large areas beneath the city are characterised by loose sand, gravel and silt, all of which are highly susceptible to liquefaction. As a result of this, the changes to the Building Code have been discussed in relation to the design of the foundations and the need to incorporate liquefaction considerations into the revised Building Codes.

Search terms: New Zealand Christchurch; Christchurch earthquake; liquefaction; building codes

blame 'nature' or the hazard for the disaster. Importantly, these anthropogenic influences are key factors that those involved in how the built environment is planned, designed, built, managed and upgraded, can influence by playing a more positive role in addressing disaster risk as an everyday part of their professional practice (see Chapter 8, particularly the 'Seven guiding principles' section).

5.2.1 The Natural Hazard

In the mid-twentieth century scientists embraced the theoretical model of plate tectonics, which was largely derived from the concept of continental drift. Plate tectonics was a way of explaining why some sections of the earth's crust (or lithosphere) appeared to be moving towards, away and/or alongside each other and thus causing earthquakes to occur. The lithosphere, which is the rigid outermost shell (or the crust and upper mantle), is broken up into what are known as tectonic plates. On Earth, there are seven or eight major plates (depending on how they are defined) and many minor plates (see Figure 5.3). Where these plates meet, their relative motion determines the type of boundary.

5.2.1.1 Types of Fault Boundaries

Some fault lines can occur a) where tectonic plates are moving towards each other (convergent), b) where they are pulling apart from each other (divergent), or c) where they are rubbing alongside each other (transform) (see Figure 5.4).

Transform boundaries are areas where two tectonic plates slide past one another. The spatial orientation of transform faults is typically parallel to the plate motions; however, this is not always

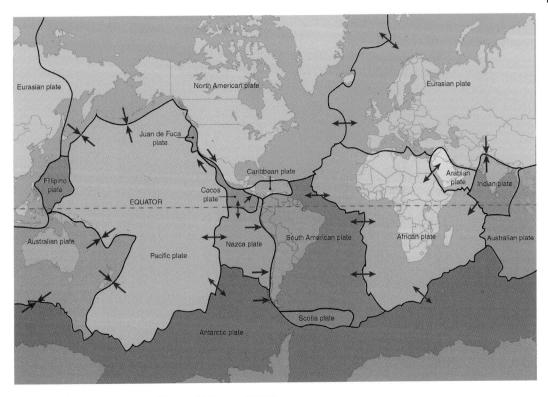

Figure 5.3 The tectonic plates of the world (*Source:* WEDC).

the case. Prominent examples of transform faults include the San Andreas Fault that runs along the Pacific coast of the USA and the North Anatolian Fault in Turkey that passes very close to Istanbul and its 14 million inhabitants.

Divergent boundaries are linear features that exist between two tectonic plates that are moving away from each other. Divergent boundaries within continents initially produce rifts which eventually become rift valleys. Most active divergent plate boundaries occur between oceanic plates and exist as mid-oceanic ridges. Divergent boundaries can also form volcanic islands which occur when the plates move apart to produce gaps which molten lava rises to fill. Examples include the Great Rift Valley in east Africa and the Mid-Atlantic Ridge that runs along the floor of the Atlantic Ocean.

Convergent boundaries are actively deforming areas where two (or more) tectonic plates or fragments of the lithosphere move towards one another and collide. Typically, the denser plate (for instance made of oceanic crust) will be subducted underneath the less dense plate (which can be either continental or oceanic crust). When both of the plates are made of oceanic crust, convergence is associated with island arcs such as the Solomon Islands. When two plates containing continental crust collide, both are too light to subduct. In this case, a continent-continent collision occurs, creating especially large mountain ranges. Examples of convergent fault boundaries include:

- **Convergent (subduction)** – Subduction of the Nazca Plate beneath the South American Plate to form the Andes
- **Convergent (continental)** – The collision between the Eurasian Plate and the Indian Plate that is forming the Himalayas

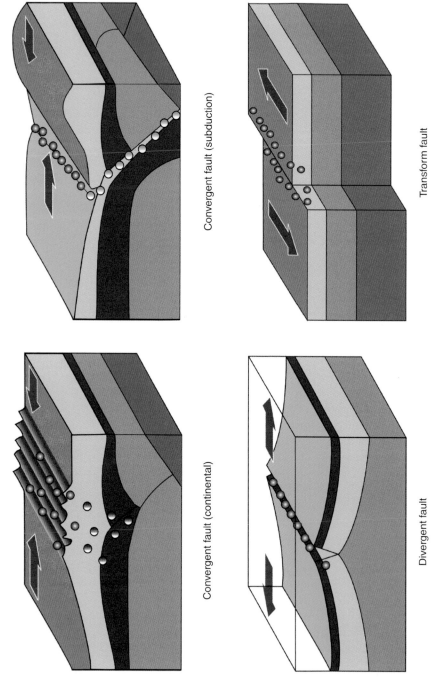

Convergent fault (subduction)

Transform fault

Convergent fault (continental)

Divergent fault

Figure 5.4 The main types of tectonic plate boundaries (*Source:* WEDC).

5.3   Seismic Activity

Seismic activity is defined "as relating to or caused by earthquakes or artificially produced earth tremors" (Collins, 2014).

Key factors:

Main causes: Earthquakes are caused by a sudden release of energy in the Earth's crust that creates seismic waves. Typically, earthquakes occur anywhere on the earth where there is sufficient stored elastic strain energy to drive fracture propagation along a tectonic fault plane. Earthquakes are influenced by proximity to, and types of, tectonic plate boundaries and the geological composition of the ground. Recent evidence also suggests that industrial activities such as 'Hydraulic fracturing', commonly termed 'fracking', can also lead to increased seismic activity.

Characteristics: Earthquakes create powerful and distinct types of seismic waves, which travel through rock with different velocities, these waves (see Figure 5.5) are categorised as:

   Longitudinal P-waves (shock- or pressure waves)
   Transverse S-waves (body waves)
   Surface waves (Rayleigh and Love waves)

The magnitude of a quake is conventionally reported by numbers on the Moment Magnitude Scale (often erroneously reported as the now superseded 'Richter scale'), whereas the felt magnitude is reported using the modified Mercalli intensity scale (intensity II–XII). The moment magnitude scale is based on the total moment release of the earthquake. Moment is a product of the distance a fault moved and the force required to move it. Magnitudes are based on a logarithmic scale (base 10). What this means is that for each whole number you go up on the magnitude scale, the amplitude of the ground motion recorded by a seismograph goes up 10 times.

Key impacts: Earthquakes can result in extensive damages that can take months (or even years) to recover from. Buildings and critical infrastructure can be damaged or destroyed, and it can become extremely difficult for affected areas to function for a prolonged period; accordingly economic losses can be vast. The impacts are further exacerbated when the seismic activity generates a tsunami (see Box 5.1).

Location: Seismologists have a reasonably good idea of where earthquakes can occur based upon records of previous seismic activity and known proximately to active faults. Substantial research has been undertaken to develop accurate seismic maps, and such maps can give a quick indication of what regions of the world are susceptible to seismic activity. However, knowing the possible location is one thing, knowing when earthquakes may strike is another more complex matter.

Advance warnings: Earthquake early warning systems have become a prime topic for research in recent decades. By drawing upon earthquake science and advancements in the technologies of monitoring systems it has been possible to develop some early warning systems (for instance, in the USA and Japan) designed to alert devices and people when shaking waves (or an associated tsunami) generated by an earthquake are expected to arrive at their location. The seconds to minutes of advance warning that such systems may provide can allow people and critical infrastructure systems to take actions to protect life and property from destructive shaking. Of course, the lead in time for these types of warnings will be highly dependent upon the proximity to the source of the seismic activity, namely, the epicentre on the Earth's surface where the seismic activity is initially monitored. The development of warning systems for tsunamis has received much investment in the last decade, especially in light of the impacts of the South Asian tsunami (2004) and

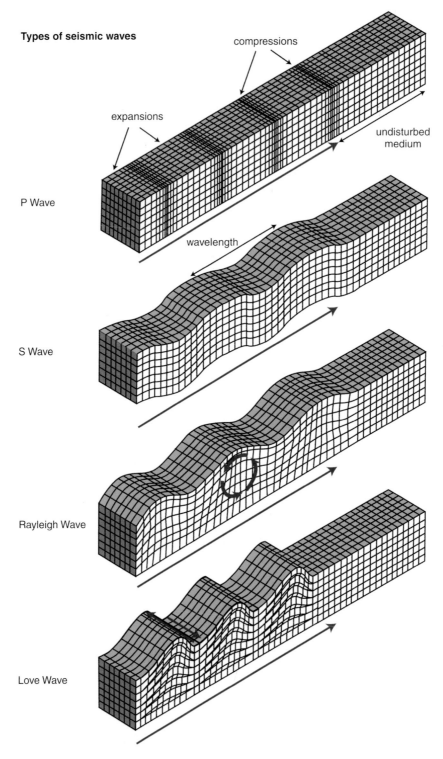

Figure 5.5 Types of seismic waves (*Source:* WEDC).

Box 5.1 What Is a Tsunami?

A tsunami (Japanese for "harbor wave"), also known as a seismic sea wave, is a series of waves in a water body caused by the displacement of a large volume of water. About 80% of tsunamis occur in the Pacific Ocean, but they are possible wherever there are large bodies of water, including lakes. Earthquakes (see Figure 5.6), volcanic eruptions (Figure 5.7), landslides (Figure 5.8), meteorite impacts and other disturbances above or below water all have the potential to generate a tsunami.

EM-DAT records show that in the 50 years between 1964 and 2013 there were 33 tsunamis globally that killed over 251,000 people, affected nearly 3 million people and caused an estimated US$222bn in damages.

Tsunami averages per year between 1964 and 2013

Major Tsunamis	0.66
Deaths	5,030
Total people affected	58,822
Total economic damages (US$)	4.4bn

---

the Tōhoku tsunami in Japan (2011). There are two distinct types of tsunami warning systems: international and regional. When operating, seismic alerts are used to instigate the watches and warnings; then, data from observed sea level height (from either shore-based tide gauges or buoys) are used to verify the existence of a tsunami.

5.4 Risk Management

5.4.1 Historical Approaches

In the past societies have tended to be aware of their natural environment and planned their settlements and built their homes and infrastructure in a way that minimises any disruptions from natural hazards. Historical records are testament to ancient civilisations suffering from the impacts of earthquakes, such as Antioch (AD115 in modern-day Syria), Lydia (AD17 in Turkey) and Crete (AD365 in Greece). However, evidence also suggests that in areas of long-term seismic activity some important buildings were designed to cope with the impacts of earthquakes. Some of these adaptations have included:

- structures designed using the principles of 'base isolation' (i.e., the Tomb of Cyrus, built during the fourth century BCE and now part of the Pasargadae World Heritage Site in Iran, see Figure 5.9).
- vernacular houses using timber-framed design with masonry infill walls (such as the *Dhajji Dewari* houses of Kashmir, see Figure 5.10 and Case Study 5.4).
- traditional houses built using a circular plan design, for instance, the *Bunga* type houses found in Gujarat, India. The walls are made of mud applied to an inner wooden trellis, a comparatively ductile and light system (see Figure 5.11).
- the use of rocking systems and segmental (multi-block) construction, for instance, in the construction of pillars in ancient Greek and Roman temples axial load for re-centring and the use of a lead element as a shear and torsion key.

5.5 Hazard Identification

Seismologists have a reasonably good idea of where earthquakes can occur based upon records of previous seismic activity and known proximately to active faults. Substantial research has been

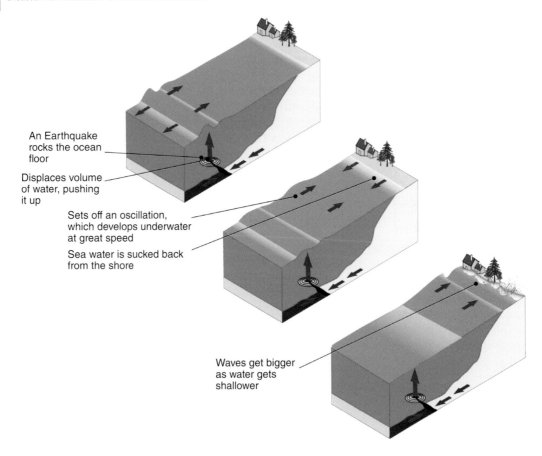

Figure 5.6 Tsunami generated by an earthquake under the sea (*Source:* WEDC).

undertaken to develop accurate seismic maps and thus such maps can give a quick indication of what regions of the world are susceptible to seismic activity. However, you do not need to be a scientist, specifically a seismologist, to understand where earthquakes can occur; a number of hazard identification methods are summarised in Table 5.3.

5.6 Assessment of the Vulnerabilities

Once an understanding has been obtained of the extent to which seismic hazards are prevalent, it is then critical to gain an appreciation of what types of sites, buildings, facilities and services could be affected by these hazards. This involves an assessment of the potential vulnerabilities of specific aspects of such locations that can be categorised as follows (and previously detailed in Chapter 2):

1) Physical vulnerability
2) Social vulnerability
3) Economic vulnerability
4) Environmental vulnerability
5) Governance vulnerability

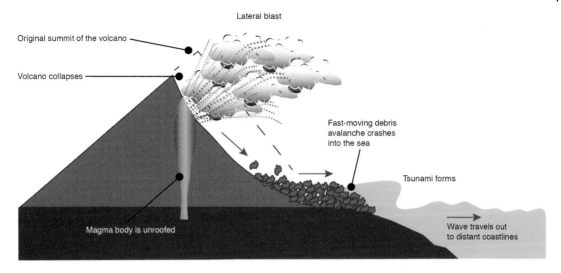

Figure 5.7 Tsunami generated by a volcano (*Source:* WEDC).

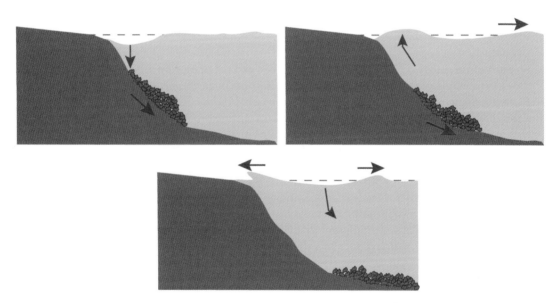

Figure 5.8 Tsunami (seiche) generated by a landslide (*Source:* WEDC).

1. **Physical Vulnerability** may be determined by aspects such as population density levels, the location (in relation to fault lines) and gradient of the site, design and materials used for critical infrastructure and for housing.
Example: Timber-framed homes are less likely to collapse in an earthquake, but they can be more vulnerable to the impacts of flash floods and coastal floods.

Figure 5.9 Tomb of Cyrus in Iran is said to be the oldest base-isolated structure in the world (*Source:* Reproduced with permission of Fahimeh Fotouhi).

2. **Social Vulnerability** refers to the root causes that make it harder for some people (i.e., women, the elderly, ethnic groups) to withstand adverse impacts of earthquake-related hazards.
   Example: Some people have limited choices about what types of buildings they inhabit, especially if they have migrated to large cities, and may end up renting rooms in over-crowded and low quality buildings; thus, when an earthquake occurs low quality buildings in densely populated areas may be especially susceptible to collapse and the knock on hazards, such as fires. These vulnerabilities can be amplified if the local inhabitants do not have land tenure rights and thus little incentive to invest in their homes in the longer term.
3. **Economic Vulnerability.** Levels of vulnerability can be highly dependent upon the economic status of individuals, communities and nations.
   Example: There is often an assumption that it can be extremely expensive to construct buildings and infrastructure to seismic codes but this is not necessarily the case. Nonetheless, for many people, even a minor increase in capital costs can be beyond their financial means.
4. **Environmental Vulnerability.** Natural resource depletion and resource degradation are key aspects of environmental vulnerability.
   Example: The overriding influence of plate tectonics suggests that earthquakes are likely to occur, irrespective of the environmental degradation that is occurring on the surface of the planet. However, it is clear that problems such as commercial deforestation can have detrimental impacts on related hazards such as landslides and also the sustainable availability of local building materials, such as trees to build timber framed houses. Recent evidence also suggests that industrial activities such as 'hydraulic fracturing' (fracking), a process used to break apart rock underground and release gas, can also lead to increased local seismic activity.

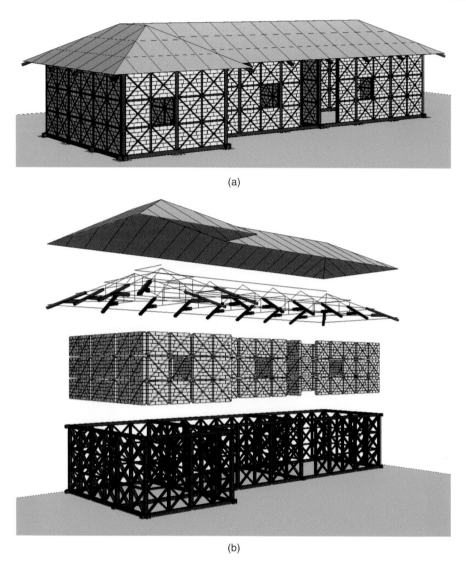

(a)

(b)

Figure 5.10 An idealised drawing of a Dhajji Dewari style building, shown in complete (a) and expanded views (b). Of course in the true vernacular form each and every Dhajji Dewari building is unique and not built to a standardised format (*Source:* Reproduced with permission of Kubilay Hicyilmaz, Arup).

5. **Governance Vulnerability.** The complex roles of national and local government in producing disaster risk and also in being affected by those risks can too often become overlooked.
   Example: Recent earthquake events have highlighted the important role of governance in determining the extent to which an earthquake hazard does or does not end up becoming a large scale disaster. The Haiti earthquake (7.0Mw) in 2010 (see Case Study 5.2) was particularly devastating due to the extensive damages caused to the built environment. One of the main reasons for such destruction was low-quality building stock and lack of enforced building standards. Structures were often informally constructed in an ad-hoc manner and some buildings were built on slopes with insufficient foundations or steel supports. In contrast, the Chilean (Maule) earthquake that occurred one month

after the Haiti earthquake, was a higher magnitude (8.8.Mw) event but it killed far less people (525 deaths in Chile compared to approximately 160,000-200,000 deaths in Haiti). This significant difference is commonly attributed to more sophisticated building codes in Chile that incorporate seismic design and the historic enforcement of those codes. It should also be noted that a mitigating factor in the Chilean earthquake was that its focus was 21 miles (34km) underground, off-shore and its epicentre was 70 miles (115km) from the nearest big city, Concepcion. The Haitian quake on the other hand was only 8 miles (13km) underground with an epicentre right on the edge of Port-au-Prince. The energy from earthquakes falls the further away you are from the centre.

Figure 5.11 Bunga style house in a village near Bhuj in Gujarat, India (*Source:* Reproduced with permission of Matthias Beckh 2001).

Case Study 5.4

Traditional Earthquake-Resistant Structures in the Himalayas

Kashmir is located in the Western part of the Himalayan mountain range on the site of a prehistoric lake created by the uplift of the mountains, therefore, making the areas rich agriculturally but also vulnerable to earthquakes.

The danger of earthquakes and soft building ground has had a great influence on the way traditional houses have been built in this region. Vernacular architecture that is adapted to local climate, culture and natural environment is one of the most prominent features of cities and villages in Kashmir. The combination of soft soils with earthquakes requires buildings that can undergo a certain amount of inelastic deformation without losing their vertical load-carrying capacity, with rigidity in construction carrying the potential for destruction. Randolph Lagenbach's excellent book '*Don't tear it down*' (2009) highlights that there are two main traditional construction systems that have proven to be effective in surviving the impacts of earthquakes: *taq* (timber-laced masonry) and *dhajji dewari* (timber frame with masonry infill).

Taq is a bearing wall masonry construction with horizontal timber lacing embedded into the masonry to keep it from spreading and cracking. One of its most unusual elements is the existence of a deliberately unbonded butt joint between the masonry piers and the wall and window panels. Although it may seem that it would weaken the building wall during an earthquake, this feature most likely helps by avoiding diagonal tension cracks from differential settlement of the foundations on the soft soil.

Dhajji dewari is a variation of a mixed timber and masonry construction type that has probably evolved due to its economic and efficient use of materials. This type of construction consists of a complete timber frame that is integral with the masonry, which fills in the openings in the frame to form walls. The frames are usually 'platform' frames with each floor being frames separately on the one below. Dhajjii dewari is very effective in holding buildings together even when they are dramatically 'out of plumb'. Structurally, the timber frame and the masonry are integral with each other; in the earthquake, the building depends on both of these elements because their interaction with each other is an important part of their ability to resist collapse (See Figure 5.10).

These traditional construction systems are frequently used together: dhajji dewari is used for the upper floors of the buildings because it is light, and taq is used on the lower floors. Dhajji dewari was also sometimes used for the back and side walls of the building with taq bearing wall construction on the front wall only. This architecture, however, is rapidly being displaced by reinforced concrete buildings because many which are constructed in a way that has proven to be dangerous in earthquakes. However, it is worth highlighting that the rather poor performance of traditional timber framed buildings in the 2015 Gorkha earthquake (Nepal) indicates that these types of traditional constructions should still be treated with caution, particularly when considering the long term and often intangible impacts of poor construction and ineffective maintenance.

Search words: vernacular architecture; traditional architecture; timer-laced masonry; timber frame; earthquake resistant traditional construction; taq; dhajji dewari

Table 5.3 Overview of Typical Earthquake Hazard Identification Methods.

Context	Hazard identification methods
Historical experience	We know where earthquakes can strike because certain locations have been affected in the (recent) past and possibly the magnitude of the earthquake was recorded by a network of seismographs. These incidents might have been recorded and reported by national or local geological survey institutions such as the British Geological Survey (BGS) in the UK, the Geological Survey of Japan (GSJ) or the U.S. Geological Survey (USGS) that have compiled detailed records of previous earthquakes and also have the capacity to monitor real time seismic activity (see below).

(Continued)

Table 5.3 (Continued)

Context	Hazard identification methods
Geological mapping	Scientific advances in the last few centuries have contributed towards the development of high resolution seismic mapping techniques and monitoring. Seismic shaking maps are prepared using consensus information on historical earthquakes, faults, and geologic materials; all this information can be combined to calculate the shaking hazard at peak ground acceleration. The potential levels of ground shaking can be used primarily for formulating building codes and for designing buildings and infrastructure. These maps can also be used for estimating potential economic losses and preparing for emergency response. It should be noted that these shaking hazard maps do not always include hazards linked to ground deformation such as liquefaction, landslides, or surface fault ruptures.
Monitoring and warning	A typical earthquake monitoring and warning system can be made up of a system of accelerometers, communication, computers, and alarms that are designed to notify local agencies of a substantial earthquake when it begins and while it is in progress. This is not the same as earthquake prediction, which is currently incapable of producing decisive event warnings (see below).
Prediction	Earthquake prediction is the ultimate goal of seismologists. Being able to predict when and where an earthquake will occur could save thousands, if not hundreds of thousands, of lives. Even after decades of study, earthquake prediction remains notoriously difficult, however, there has been an increasing interest in the identification or seismic 'precursors' that may act as early indicators of an earthquake. For instance, some scientists have posited (but not yet really proven) that prior to an earthquake there can be cases of abnormal behaviour patterns of animals (such as birds and dogs), and/or other phenomena such as the swift release of ozone gas from different igneous and metamorphic rocks such as basalt, granite, rhyolite, gneiss and quartz; more work has yet to be undertaken to ascertain whether such phenomena may be useful in providing early warning of an imminent earthquake.

5.7 Determination of the Risk

Effects associated with earthquakes can be divided into primary and secondary. The primary effects of earthquakes are those caused by the actual shaking of the ground resulting in damage to buildings, infrastructure, and even in the case of liquefaction the loss of stable land. The secondary effects can be short term (tsunamis and landslides) or longer-term effects that are indirectly related to earthquakes (see Table 5.4 for more examples). Scientists have developed models designed to estimate damage and loss of buildings, lifelines and essential facilities from scenario and probabilistic earthquakes, including:

- Ground shaking and ground failure
- Estimates of casualties
- Displaced households and shelter requirements
- Damage and loss of use of essential facilities
- Estimated cost of repairing damaged buildings
- Quantity of debris
- Damage to buildings
- Direct costs associated with loss of function (e.g., loss of business revenue)

5.7.1 Earthquake Damage Estimation

Earthquake damage estimation can be a very useful exercise in the management of seismic risk by informing the decision-making process. Hazus is a standardized methodology used in the USA that contains models for estimating potential losses from earthquakes (as well as floods and hurricanes).

Table 5.4 Examples of the Primary and Secondary Effects of Earthquakes (Excluding Tsunamis).

	Primary effects	Secondary effects
Physical	*Damage to...* • Critical infrastructure • Houses • Offices/shops • Schools • Vehicles • Equipment • Agriculture	*Including fires, tsunamis and landslides, resulting in...* • Homelessness/displacement • Disrupted essential services • Reduced access for heavy plant/vehicles • Leaking/contaminated water channels/sewers • Polluted land (due to salinization of soil) • Disrupted education provision • Loss of crops
Wellbeing/health	*Deaths or injuries to...* • People • Livestock	*Leading to...* • Post-traumatic stress disorder • General fear of safety due to aftershocks • Unhygienic conditions • Unsafe land/homes • Polluted land/homes • Strains upon social support services
Economic	*Immediate...* • Repair/replacement costs • Loss of business	*Longer term...* • Rehabilitation costs • Insurance costs (and increased premiums) • Reduced commerce • Loss of crops/yields • Loss of tourist revenues • Reduced land values

Hazus uses Geographic Information Systems (GIS) technology to estimate physical, economic and social impacts of disasters. It graphically illustrates the limits of identified high-risk locations due to anticipated earthquake scenarios. Users can then visualise the spatial relationships between populations and other more permanently fixed geographic assets or resources for the specific hazard being modelled, a crucial function in the pre-disaster planning process. There are a number of other earthquake damage methodologies but it is fair to say that Hazus is considered to be one of the most valuable tools has it has been regularly refined since its initial release in 1997 by the Federal Emergency Management Agency (FEMA).

Figure 5.12 provides an illustration of how the Hazus tool can be used to better understand the relationship utility lifeline system components and the earthquake hazard in Portland, based on soil liquefaction potential. This tool can be used in pre-disaster context for the:

• Evaluation of options for future extension of utility lifeline systems in Portland in relation to areas of potential soil liquefaction.
• Prioritization of mitigation measures for system components to reduce vulnerability to potential liquefaction.
• Evaluation of the population and businesses that are potentially vulnerable to the earthquake hazard in Portland.

Also, the same tool can be useful in a post-disaster context for the:

• Identification of utility lifelines that are potentially impacted by soil liquefaction.
• Identification of utilities that are potentially eligible for funds under FEMA's Public Assistance program following a declared disaster.
• Identification of mitigation opportunities following a damaging earthquake.

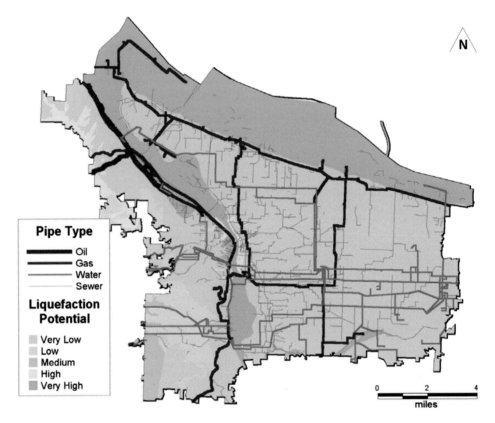

Figure 5.12 Distribution of utility lifelines in relation to earthquake hazard in Portland, Oregon, USA (*Source:* Reproduced with permission of DHS).

5.8 Identification and Prioritisation of Risk Reduction Options

As discussed in Chapter 2, there is a number of ways to reduce the risk of a hazard. However, in the case of earthquakes there are fewer options available when compared to dealing with other hazards such as floods (see Chapter 3). As outlined in Chapter 2, the best approach to considering risk reduction measures should adopt the five interrelated stages as shown in Table 5.5, which summarises the extent to which each of the risk reduction options can be utilised for earthquakes and associated tsunamis. A more detailed list of specific risk reduction examples is provided for dealing with seismic risk in Table 5.6.

Knowledge of local amplification of the seismic motion from the bedrock is very important in order to choose the most suitable design solutions. Local amplification can be anticipated from the presence of particular stratigraphic conditions, such as soft soil overlapping the bedrock, or where morphological settings (e.g., steep slopes, valleys, or drainage basins) may produce a concentration of the seismic event. The identification of the areas potentially affected by earthquake-induced landslides and by soil liquefaction can be made by geological survey and by analysis of historical documents. Areas prone to the impacts of tsunamis can also be relatively easy to identify.

Table 5.5 Summary of the Viability of Risk Reduction Options for Addressing Earthquake (and Associated Tsunami) Risks.

Type of risk reduction option for earthquake risk					
Inherent safety	Prevention	Detection	Control	Mitigation	Emergency response
N	N	N	N	Y	Y
Type of risk reduction option for tsunami risk					
Inherent safety	Prevention	Detection	Control	Mitigation	Emergency response
N	N	Y	N	#	#

Please note:
'Y' – indicates that there are possibly a range of useful risk options available
'#' – indicates that some risk reduction options can be used but they are likely to be of only limited effectiveness
'N' – indicates that other than relocating the built asset there is little that can be done to reduce this hazard/threat

Table 5.6 Indicative Examples of Risk Reduction Options for Addressing Earthquake Risk.

Risk reduction option	Examples
1) Inherent safety – eliminate the possibility of hazards occurring	Not applicable
2) Prevention – reduce the likelihood of hazards	Not applicable
3) Detection – measures for early warning of hazards	Not yet sufficiently effective for earthquakes Tsunami monitoring and warning systems can be used (i.e., in the Pacific).
4) Control – limiting the size of the hazards	Not applicable. It may be possible to design buildings that reduce the impacts, but this action would then fall under 'Mitigation and adaptation'.
5) Mitigation and adaptation – protection from the effects of hazards	Adopting and enforcing updated building code provisions to reduce earthquake damage risk. Bear in mind that the enforcement of such codes can be problematic. Earthquake risk can be reduced through local planning, that is, avoiding areas prone to liquefaction. Conducting seismic retrofitting for critical public facilities most at risk to earthquakes. Large walls can be built to mitigate the impact of tsunamis but the success of these types of expensive and intrusive structures has been limited. Conduct inspections of building safety (i.e., government buildings, hospitals, schools and other critical infrastructure). Using flexible piping when extending water, sewer, or natural gas services Building a safe room to provide protection during an earthquake. Developing an outreach program about earthquake risk and mitigation activities in homes, schools, and businesses. Educating homeowners about structural and non-structural retrofitting of vulnerable homes and encouraging retrofit.

(Continued)

Table 5.6 (Continued)

Risk reduction option	Examples
6) Emergency response – planning for evacuation emergency access	Draw up detailed vulnerability maps. This could include developing an inventory of public and commercial buildings assets that may be particularly vulnerable to earthquake damage. Ensure that the emergency services stations and facilities are designed to suitable seismic codes. Identify safe evacuation and emergency access routes.

5.8.1 Inherent Safety

The only way that this could be achieved would be to avoid building (or indeed demolished existing developments) on land that is prone to seismic activity. This approach may, in theory rather than practice, be possible for undeveloped land/regions, but in the twenty-first century we are dealing with a legacy of human habitations that have been built in seismically active areas so this option is not really viable.

5.8.2 Detection of Hazard

It may be possible to detect earthquakes but at the moment the science behind the detection systems provides us with very limited benefits. Earthquake early warning systems have become a prime topic for research in recent decades. The seconds to minutes of advance warning that such systems may provide can allow people and critical infrastructure systems to take actions to protect life and property from destructive shaking. However, the lead in time for these types of warnings will be highly dependent upon the proximity to the source of the seismic activity, namely the epicentre on the Earth's surface where the seismic activity is initially monitored.

5.8.3 Mitigation of Hazard

Seismic analysis is a subset of structural analysis and is the calculation of the response of a building or other structure to earthquakes. It is part of the process of structural design, earthquake engineering or structural assessment and retrofit in regions where earthquakes are prevalent. In essence, a building has the potential to 'wave' back and forth during an earthquake (or even during a severe wind). This is called the 'fundamental mode', and is the lowest frequency of building response. Most buildings, however, have higher modes of response, which are uniquely activated during earthquakes.

Earthquake engineering has developed a lot since the early days, and some of the more complex designs now use special earthquake protective elements either just in the foundation (base isolation) or distributed throughout the structure. Analysing these types of structures requires specialised explicit finite element computer code, which divides time into very small slices and models the actual physics, much like common video games often have 'physics engines'.

5.8.4 Earthquake-Resistant Construction

Earthquake construction means implementation of seismic design to enable building and non-building structures to live through the anticipated earthquake exposure up to the expectations and in compliance with the applicable building codes.

Design and construction are closely related. Central to this is the need for good workmanship and thus the detailing of the members and their connections should be as simple as possible. As any construction in general, earthquake construction is a process that consists of the building, retrofitting or assembling of infrastructure given the construction materials available. The crucial fact is that, for safety, earthquake-resistant construction techniques are as important as quality control and using correct materials. Each construction project requires a qualified team of professionals who understand the basic features of seismic performance of different structures as well as construction management. A summary table of some of the key earthquake engineering considerations for different types of construction design are provided in Table 5.7. Specific types of earthquake engineering designs are highlighted in Box 5.2.

5.8.5 Mitigation of Tsunamis

Japan, where tsunami science and response measures first began following the Sanriku earthquake in 1896, has produced ever-more elaborate countermeasures and response plans. That country has built many tsunami walls of up to 12 metres (39ft) high to protect populated coastal areas. Other

Table 5.7 Overview of Most Prominent Construction Types That Can Incorporate Earthquake Engineering Features.

Type of construction	Key considerations
Adobe structures	Around one-third of the world's population lives or works in buildings made out of earth. Adobe buildings consisting of mud bricks are some of the oldest and most widely used types of buildings. The use of adobe is very common in some of the world's most hazard-prone regions, traditionally across Latin America, Africa, Indian subcontinent and other parts of Asia, Middle East and Southern Europe. Adobe buildings are considered particularly vulnerable to earthquakes. However, there are number ways of that new and existing adobe buildings can be strengthened. Key factors for the improved seismic performance of adobe construction are: • Quality of construction; good workmanship and use of natural fibres (such as straw to reinforce the clay/brick materials). • Compact, robust layout, single storey with small openings. • Seismic features such as buttresses and use of cane for reinforcement
Limestone and sandstone structures	Limestone is very common in architecture, especially in North America and Europe. Many medieval churches and castles in Europe are made of limestone and sandstone masonry. While these are durable materials their rather heavy weight is not necessarily beneficial for adequate seismic performance. Application of new technologies to seismic retrofitting can enhance the survivability of unreinforced masonry structures. The Utah State Capitol Building in Salt Lake City is a good example of what can be achieved; in the 1980s, it became the first building in the world to be retrofitted with base isolators. During an earthquake, the building will gently float on 440 steel and rubber pads (see Figure 5.13). It is also commendable that the building was exhaustively renovated and repaired with an emphasis on preserving the historical accuracy in appearance.
Timber frame structures	Timber-framed structures date back thousands of years, and have been used in locations as diverse as ancient Japan, Europe and medieval England; typically in regions where timber was in ample supply and building stone and the skills to work it were not available or were considered prohibitively expensive. The use of timber framing in buildings provides skeletal framing which if properly engineered, lends itself to better seismic survivability. The traditional *dhajji dewari* style houses (see Case Study 5.4) found in mountainous regions of South Asia are good examples of vernacular building designs that have evolved over time to incorporate sound earthquake engineering principles.

(Continued)

Table 5.7 (Continued)

Type of construction	Key considerations
Reinforced masonry structures	The use of plain or unreinforced masonry is limited in scope and structural capacity which is needed for large building systems, or when the lateral forces due to wind and earthquakes are significantly larger. Use of reinforcement in masonry enhances its flexure and shear capacity which can overcome the limitations in the use of unreinforced masonry. Reinforced masonry is a construction system where steel reinforcements in the form of reinforcing bars or mesh are embedded in the mortar or placed in the holes and filled with concrete or grout. By reinforcing the masonry with steel reinforcement, the resistance to seismic loads and energy dissipation capacity can be improved significantly. There are also examples of unconventional approaches to the reinforcement of masonry structures. For instance, timber-reinforced stone masonry (known as *Koti Banal* Architecture) of Uttarakhand and Himachal Pradesh, in Northern India, is an impressive approach to building multi-story buildings in seismically active areas. Koti Banal structures (see Figure 5.14) in general have a single small entry point and relatively small openings, which are surrounded by strong wooden elements to compensate for the loss of strength; no windows are provided at ground floor level.
Reinforced concrete structures	Reinforced concrete is concrete in which steel reinforcement bars (rebars) or fibres have been incorporated to strengthen a material that would otherwise be brittle. Pre-stressed concrete is a kind of reinforced concrete used for overcoming concrete's natural weakness in tension. It can be applied to beams, floors or bridges with a longer span than is practical with ordinary reinforced concrete. Pre-stressing tendons (generally of high tensile steel cable or rods) are used to provide a clamping load which produces a compressive stress that offsets the tensile stress that the concrete compression member would, otherwise, experience due to a bending load. To prevent catastrophic collapse in response earth shaking (in the interest of life safety), a traditional reinforced concrete frame should have ductile joints. Depending upon the methods used and the imposed seismic forces, the post-earthquake condition of such buildings may require extensive repair, or may have to be demolished.
Pre-stressed structures	Pre-stressing involves the creation of permanent stresses in a structure for the purpose of improving its performance under various service conditions. There are a number of basic types of pre-stressing: <ul><li>Pre-compression (mostly, with the own weight of a structure)</li><li>Pre-tensioning with high-strength embedded tendons</li><li>Post-tensioning with high-strength bonded or unbonded tendons</li></ul> The concept of a pre-stressed structure (a technique used by the ancient Romans) is now generally utilised in the design of buildings, underground structures, TV towers, power stations, floating storage and offshore facilities, nuclear reactor vessels, and numerous kinds of bridge systems.
Steel structures	Steel structures have often been considered to perform well in resisting the impacts of seismic activity, but some failures have occurred. A significant number of welded steel moment-resisting frame buildings experienced brittle behaviour and were perilously damaged in the 1994 Northridge earthquake in California. After that, the Federal Emergency Management Agency initiated the development of repair techniques and new design approaches to minimize damage to steel moment frame buildings in future earthquakes. As a consequence of the Northridge earthquake experience, the American Institute of Steel Construction introduced AISC 358 'Pre-Qualified Connections for Special and intermediate Steel Moment Frames'. The AISC Seismic Design Provisions require that all Steel Moment Resisting Frames employ either connections contained in AISC 358, or the use of connections that have been subjected to pre-qualifying cyclic testing.

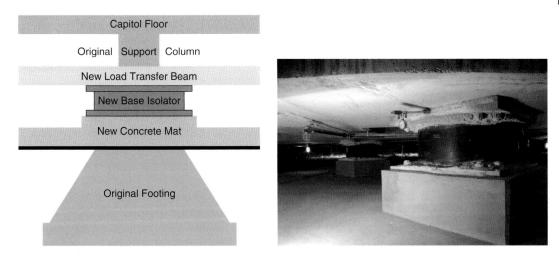

Figure 5.13 (left) A cross-section of the base isolator system; (right) A base isolator under the Utah State Capital (*Source:* Lee Bosher).

Figure 5.14 Example of a Koti Banal (timber reinforced masonry) building in Northern India (*Source:* Reproduced with permission of Dominik Lang).

localities have built floodgates of up to 15.5 metres (51ft) high and channels to redirect the water from the incoming tsunami. However, the effectiveness of these types of structures has been questioned, as too often tsunamis have managed to overtop the barriers. While tsunami defensive walls may not have been entirely successful, the construction of strategically located tsunami shelters

---

Box 5.2 Some Key Earthquake Engineering Approaches to Improve Seismic Performance.

Base isolation is one of the most popular means of protecting a structure against earthquake forces. Figure 5.9 shows the Tomb of Cyrus in Iran, which is considered to one of the earliest surviving examples of base isolation. It is a collection of structural elements which effectively decouple a superstructure from its substructure resting on a shaking ground thus protecting a building's integrity. The City Hall in Los Angeles, USA, is one of the tallest base isolated buildings in the world.

Tuned mass damper, also known as a harmonic absorber, is a device mounted in structures to reduce the amplitude of mechanical vibrations such as seismic waves. Their application can prevent discomfort, damage, or outright structural failure. The Taipei 101 skyscraper in Taiwan contains the world's largest and heaviest tuned mass dampers, weighing in at 660 metric tons (see Figure 5.15).

Vibration control, in earthquake engineering, refers to the technical measures aimed to mitigate seismic impacts in buildings and other structures. These measures work by a) dissipating the wave energy inside a superstructure with suitably engineered dampers; b) dispersing the wave energy between a wider range of frequencies; and/or c) by absorbing the resonant portions of the whole wave frequencies band with the help of mass dampers.

All seismic vibration control devices may be classified as passive, active or hybrid where:

- *passive control* devices have no feedback capability between them, structural elements and the ground;
- *active control* devices incorporate real-time recording instrumentation on the ground integrated with earthquake input processing equipment and motors within the structure;
- *hybrid control* devices have combined features of active and passive control systems.

The types of options that are most suitable to use are typically based on very context specific circumstances such as ground conditions and proximity to fault lines. Therefore, it is important to seek expert advice on all aspects of earthquake engineering before deciding on what hazard mitigation measures should be used. This guide is merely that; an introduction to the range of options that can be utilised.

(see Figure 5.16) can provide essential for the vertical evacuation of tsunami prone communities, especially in low-lying coastal areas where horizontal evacuation may be unfeasible.

5.8.6 Emergency Response

Preparedness planning, as it is widely acknowledged how difficult, arguably impossible, it is to provide sufficiently long warnings for earthquakes, thus preparedness planning plays an especially important role in enabling affected communities to cope with and recover from damaging seismic activity. The American Red Cross recommends that the general public and businesses undertake the following activities when preparing for earthquakes:

- Become aware of fire evacuation and earthquake plans for all of the buildings you occupy regularly (see Figure 5.17).
- Pick safe places in each room of your home, workplace and/or school. A safe place could be under a piece of furniture or against an interior wall away from windows, bookcases or tall furniture that could fall on you.
- Practice drop, cover and hold on in each safe place. If you do not have sturdy furniture to hold on to, sit on the floor next to an interior wall and cover your head and neck with your arms.
- Keep a flashlight and sturdy shoes by each person's bed.

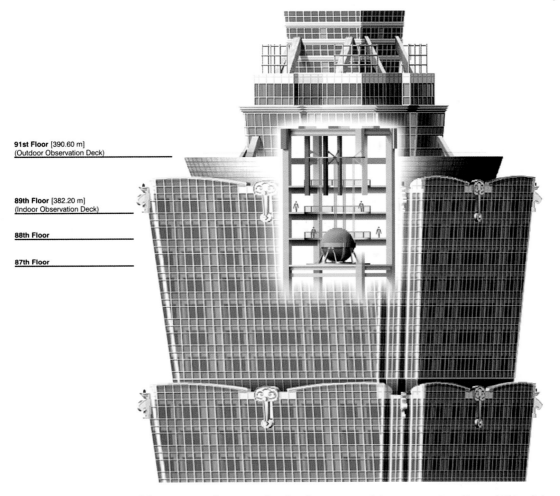

91st Floor [390.60 m]
(Outdoor Observation Deck)

89th Floor [382.20 m]
(Indoor Observation Deck)

88th Floor

87th Floor

Figure 5.15 Cross section of the Taipei 101 skyscraper showing the mass tuned dampener system (*Source:* Wikipedia).

- Make sure your home is securely anchored to its foundation.
- Bolt and brace water heaters and gas appliances to wall studs.
- Bolt bookcases, china cabinets and other tall furniture to wall studs.
- Hang heavy items, such as pictures and mirrors, away from beds, couches and anywhere people sleep or sit.
- Brace overhead light fixtures.
- Install strong latches or bolts on cabinets. Large or heavy items should be closest to the floor.
- Learn how to shut off the gas valves in your home and keep a wrench handy for that purpose.
- Learn about your area's seismic building standards and land use codes before you begin new construction.
- Keep and maintain an emergency supplies kit in an easy-to-access location.

Draw up detailed vulnerability maps. This activity could include local authorities, communities or businesses developing an inventory of public and commercial buildings assets that may be particularly vulnerable to earthquake damage. A key part of this approach is ensuring that the emergency services stations and facilities are designed to suitable seismic codes, as well as (ideally when feasible) the transport routes to and from those facilities.

Identify safe evacuation and emergency access routes can be of particular relevance when considering the risk of tsunamis where a warning of even a few minutes can save lives. Evacuation routes can be horizontal where the local topography enables evacuation to nearby hills of 10 meters or higher. In extremely low lying coastal areas, it may be more pragmatic to construct tsunami shelters (Figure 5.16) so that vertical evacuation can be made possible. Of course, any evacuation procedures are highly reliant on effective tsunami monitoring and warning systems.

5.9   Summary

There is increasing acknowledgement that earthquakes are natural processes that become disastrous events due to human activities, such as:

- Poor (or unregulated) urban planning (building homes, offices, infrastructure and essential services in seismically active locations).
- Buildings and infrastructure not designed to cope with the physical impacts of earthquakes.
- Artificial development of land converted from wetland areas or the sea; such land can be prone to the impacts of liquefaction.
- Lack of, or ineffective, emergency preparedness and evacuation procedures.

While this may sound like a negative perspective, it is not necessarily the case. The human influence on whether an earthquake becomes a disaster or not suggests that there is actually quite a lot that we (as construction practitioners, home owners and individuals) can do to reduce the impacts of seismic activity.

Figure 5.16   Tsunami shelter in Khao Lak, Thailand (*Source:* Roy Googin).

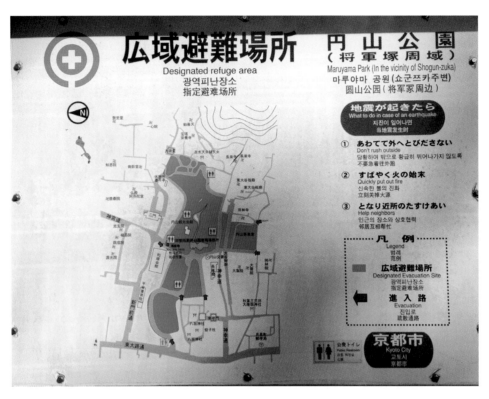

Figure 5.17 Information board showing earthquake evacuation routes and designated safe refuges in Kyoto, Japan (*Source:* Ksenia Chmutina).

Key Points

- While invariably it will be impractical, and indeed unnecessary, for all buildings and structures to be built (or retrofitted) to stringent seismic codes, it is important for such considerations to be undertaken in the most critical of cases (i.e., related to critical infrastructure, schools and hospitals).
- Holistic multi-hazard multi-stakeholder approach should be mainstreamed in order to increase resilience of the built environment
- Traditional approaches to earthquake engineering through vernacular design should not be forgotten. There are many examples in this chapter of how local materials can be used to construct non-engineered buildings that can survive earthquakes.
- It is very difficult in cost and effectiveness terms to build defences that can protect coastline from the impacts of tsunamis. Investment in tsunami monitoring, warning and evacuation systems (developed with and by local communities) can be effective in at least reducing the casualties from tsunamis.

Reinforcement

- What is the 'scale' called that earthquake magnitude is measured on?
- What are the most prominent earthquake engineering design approaches that can be used to improve seismic performance of buildings and structures?
- What is the role of construction stakeholders in DRR?
- What is the difference between structural and non-structural approaches?

Questions for discussion

- In what circumstances is it fair to state that earthquake disasters are the result of human activities?
- Why do you think it can be difficult to enforce seismic building codes?
- What are the best ways of raising awareness of what people should do in the event of an earthquake?
- Do you think earthquake engineering should be a professional competency for all civil engineers

Further Reading

Books:

Beer, M., Kougioumtzoglou I.A., Patelli E. and Au I., (2015). *Encyclopedia of earthquake engineering*, Springer

Bothara, J., and Brzev, S. (2012). *A tutorial: improving the seismic performance of stone masonry buildings*. Earthquake Engineering Research Institute, Oakland, California

Chu, S.Y.; Soong, T.T. and Reinhorn, A.M. (2005). *Active, Hybrid and Semi-Active Structural Control*, John Wiley & Sons. ISBN 0-470-01352-4

Elnashai, A.S. and Di Sarno, L., (2015). *Fundamentals of Earthquake Engineering: From Source to Fragility*, John Wiley & Sons

Govorushko, S.M., (2011). *Natural processes and human impacts: interactions between humanity and the environment*, Springer Science & Business Media,

Fardis, M. N., (2015). *Seismic design of concrete buildings to Eurocode 8*, CRC Press.

Langenbach, R., (2009). *Don't tear it down!: Preserving the earthquake resistant vernacular architecture of Kashmir*, UNESCO, published by Oinfroin Media, Oakland

Smith, K., (2013). *Environmental Hazards: Assessing Risk and Reducing Disaster*. (6th Edition), Routledge

Articles/reports:

Ambraseys, N. and Bilham, R., (2011). 'Corruption kills', *Nature*, 469(7329), pp. 153–155

Bilham, R., (2015). 'Seismology: raising Kathmandu', *Nature Geoscience*, 8(8), pp. 582–584

Bilham, R., (2010). 'Lessons from the Haiti earthquake', *Nature*, 463(7283), pp. 878–879

Lewis, J., (2008). 'The worm in the bud': in Bosher L.S. (ed.) *Hazards and the Built Environment: Attaining Built-in Resilience*, Routledge, London: p. 238

Rautela P. and Joshi G.C., (2007) *Earthquake Safe Koti Banal Architecture of Uttarakhand (India)*, Disaster Mitigation and Management Centre, Department of Disaster Management, Dehradun

Useful websites:

https://www.eeri.org/
http://www.usgs.gov/
Tsunami data and information: https://www.ngdc.noaa.gov/hazard/tsu.shtml

6

Volcanoes

Now came the dust, though still thinly. I look back: a dense cloud looms behind us, following us like a flood poured across the land. "Let us turn aside while we can still see, lest we be knocked over in the street and crushed by the crowd of our companions." We had scarcely sat down when a darkness came that was not like a moonless or cloudy night, but more like the black of closed and unlighted rooms. You could hear women lamenting, children crying, men shouting................

Pliny the Younger's Description of the 79 A.D. eruption of Vesuvius

Historical accounts and abandoned cities are testament to the impacts of volcanic activity and how these natural carbuncles have supported and blighted the development of human habitations for thousands of years (see Figure 6.1). While the impacts of volcanic eruptions can be extensive, affecting a vast geographical area, recent data suggests that volcanoes typically account for the lowest proportion of deaths (552 per year) and economic damages (US$61m per year) than the other types of geophysical or hydrometeorological hazards discussed in this book.

6.1 Learning Objectives

By the end of this chapter you will learn:

- What are the main causes of volcanic eruptions?
- What are the typical impacts volcanoes?
- How to identify volcanic risks and assess vulnerabilities?
- How to reduce the impacts of volcanic activity using structural and non-structural measures?

6.2 Living with Volcanoes

There are approximately 500 active volcanoes in the world. Historical records have documented some of the long-term fragile relationships between the establishment of human settlements in locations in close proximity to volcanoes (see Table 6.1 and Figures 6.1 and 6.2). Pliny the Younger (*Gaius Plinius Caecilius Secundus*), who was a lawyer, author, and magistrate of Ancient Rome, witnessed the eruption of Vesuvius in 79 AD, in which his uncle, Pliny the Elder, died. The eruption of Mount Vesuvius in 79 AD was one of the most catastrophic and infamous volcanic eruptions in

Disaster Risk Reduction for the Built Environment, First Edition. Lee Bosher and Ksenia Chmutina.
© 2017 John Wiley & Sons Ltd. Published 2017 by John Wiley & Sons Ltd.

Figure 6.1 Mount Calbuco erupts high above Puerto Montt on the Chilean coastline (*Source:* Reproduced with permission of Carolina Barría Kemp).

Table 6.1 The 10 Most Deadly Volcanoes.

Volcano, country	Year	Total deaths	Main causes of death
Tambora, Indonesia	1815	92,000	Starvation
Krakatau, Indonesia	1883	36,417	Tsunami
Mt. Pelee, Martinique	1902	29,025	Ash flows
Ruiz, Colombia	1985	25,000	Mudflows
Unzen, Japan	1792	14,300	Volcano collapse, tsunami
Laki, Iceland	1783	9,350	Starvation
Kelut, Indonesia	1919	5,110	Mudflows
Galunggung, Indonesia	1882	4,011	Mudflows
Vesuvius, Italy	1631	3,500	Mudflows, lava flows
Vesuvius, Italy	79	3,360	Ash flows and falls

European history. Several Roman settlements (such as Pompeii and Herculaneum) were obliterated and buried underneath massive pyroclastic surges and ashfall deposits (see Case Study **6.1**). Volcanoes not only produce various hazards, but eruptions that can last days or even years (e.g., Mt Etna, Sicily). Volcanic eruptions, therefore, have a tendency to influence culture, livelihood and reasoning at the local scale. Thus, the relationship between people and volcanoes can often create various oral traditions and mitigation actions over time; some of these will be highlighted later in this chapter.

6.3   Overview of the Typical Impacts of Volcanoes

According to records compiled by EM-DAT (see Table 6.2 for yearly averages), in the 50 years between 1964 and 2013 there were 192 significant volcanic eruptions globally, that killed over 27,000 people, affected nearly 5 million people and caused an estimated US$3bn in damages.

The primary effects of volcanoes can be extensive and diverse and will be explained in more depth later in the chapter. In essence, primarily **lava flows** can be very damaging to property as they can destroy anything in their path. Hot **pyroclastic flows** can cause death by suffocation and burning. They can travel so rapidly that few humans are able to escape, even if using motor vehicles. The **lateral blasts** can knock down substantial structures in their path, including engineered buildings and infrastructure. **Tephra (fragmental material) falls** can cause the collapse of roofs and can affect areas far away from the initial eruption. Tephra falls can also destroy vegetation, including crops, and can kill livestock that eat the ash-covered vegetation. Volcanoes may also emit **poisonous gases**, such as carbon dioxide (CO_2), and sulphur dioxide (SO_2) that can be toxic to living organisms.

The secondary effects of volcanoes can also be extensive and are typically related to:

Lahars – Lahars are mudflows and debris flows generated by volcanic activity. These can be generated when the heat from the volcano melts snow or ice during the eruption, emptying of crater lakes during an eruption, or rainfall that takes place any time with no eruption. Mudflows are a mixture of water and sediment that varies between thick water and wet aggregations, and can remove anything in their paths like bridges, highways, houses, and so on (see Figure 6.3). Debris flows can occur when the sides of volcanic mountains become very steep as a result of the addition

---

CASE STUDY 6.1

Pompeii

The city of Pompeii was an ancient Roman city near modern Naples. It was mostly buried under 4 to 6 metres (13–16 feet) of ash and pumice due to the eruption of Mount Vesuvius in 79 AD. The people and buildings of Pompeii were covered in up to 12 different layers of tephra, in total 25 meters (82 feet) deep, which rained down for about 6 hours. By the first century AD, Pompeii was one of a number of towns located near the base of the volcano, since the area had a substantial population which grew prosperous from the region's renowned agricultural fertility (see Case Study **6.2**). The city was uncovered by the exploration of the ancient site started in the Civita area in 1748. The excavations still take place today, as many areas are still to be uncovered in Pompeii, but it is even more important to restore what has already been excavated. Today, 44 of the 66 hectares of urban area are visible, and it is unanimously considered that the other 22 hectares must be left under the volcanic debris, in order to preserve this important part of our past for future generations. At the time of the destruction, the city was about 160 years old with a population of approximately 11,000 people and had a complex water system, an amphitheatre, a gymnasium and a port. Evidence for the destruction originally came from a letter to a friend by Pliny the Younger, who saw the eruption from a distance and described the death of his uncle. Numerous studies of the eruption products and victims indicate that heat – rather than suffocation – was the main cause of death of people: exposure to at least 250 °C hot surges at a distance of 10 kilometres from the vent was sufficient to cause instant death, even if people were sheltered within buildings.

Search terms: Pompeii; history of Pompeii; Vesuvius volcano

Figure 6.2 Mount Vesuvius looms large, located only 9 km (5.6 miles) east of Naples which is home to 3 million inhabitants (*Source:* Reproduced with permission of Abi Brooks).

Table 6.2 Volcano Averages Per Year Between 1964 and 2013.

Major volcanic eruptions	3.84
Deaths	552
Total people affected	98,062
Total economic damages (US$)	61m

of new material over time as well due to inflation of the mountain as magma intrudes. Overly steep slopes can result in landslides, debris slides or debris avalanches.

Flooding – Natural (such as river channels) and human-made drainage systems can become blocked by deposition of pyroclastic flows and lava flows. Such blockages may create a temporary dam that could eventually fill with water and suddenly fail resulting in floods downstream. Volcanoes in cold climates can melt snow and glacial ice, rapidly releasing water into the drainage system and possibly causing floods.

Tsunamis - Debris avalanche events, landslides, caldera collapse events, and pyroclastic flows entering a body of water may generate tsunamis (also see Chapter 5).

Volcanic Earthquakes and Tremors - Earthquakes usually precede and accompany volcanic eruptions, as magma intrudes and moves beneath and within the volcano. Although most volcanic earthquakes are small, some are large enough to cause damage in the area immediately surrounding the volcano, and some are large enough to trigger landslides and debris avalanches.

Figure 6.3 Plymouth is the capital city of the Caribbean island of Montserrat, since initial eruptions in 1995 the town has now been buried by a series of pyroclastic flows and lahars. This image shows the remains of a Catholic Church, Plymouth in February 2006. (*Source:* Reproduced with permission of Richard Roscoe).

Atmospheric Effects - Since large quantities of tephra and volcanic gases can be injected into the atmosphere, volcanism can have a short-term effect on climate. Ash in the atmosphere may ground aircraft as there is a concern that the particles can get into the engines and cause them to stall.

Famine and Disease - Ash falls can cause extensive crop damage and kill livestock and this can lead to famine. People may have to be evacuated and thus cut off from other normal services; this can lead to disease for years after an eruption, especially if the infrastructure is not in place to provide for rapid relief and recovery.

6.4 Causes of Volcanoes

Volcanoes are notorious natural hazards, and while the causes of them are well understood the ability to predict when some volcanoes will erupt is a less well-defined science. Like other natural hazards, such as floods and windstorms, there is increasing acknowledgement that volcanoes are natural processes that become disastrous events due to human activities; such as:

- A long-established urge for famers to utilise the very fertile soils that typically lie close to volcanoes (Case Study **6.2**).
- Poor (or unregulated) land use planning (building homes, offices, infrastructure and essential services in volcanically active locations).
- Lack of, or ineffective, emergency preparedness and evacuation procedures.

All of these factors are human induced and play important roles in how natural processes such as volcanic activity can lead to disasters. Importantly, these human influences are key factors that those involved in how the built environment is planned, designed, built, managed and upgraded, can address by playing a more positive role in incorporating disaster risk as an everyday part of their professional practice.

CASE STUDY 6.2

Agriculture Near Vesuvius

Mount Vesuvius is a stratovolcano in the Gulf of Naples, Italy, about 9 km east of Naples and a short distance from the coastline. It is best known for its eruption in AD 79 that led to the burying and destruction of Pompeii (see Case Study **6.1**). Vesuvius has erupted many times since and is the only volcano on the European mainland to have erupted within the last hundred years. Today, it is regarded as one of the most dangerous volcanoes in the world because of the 3 million people population living nearby and its tendency towards explosive (Plinian) eruptions. The area around Vesuvius is declared a national park and is also well known for its agricultural produce. Two large eruptions 35,000 and 12,000 years ago left the region blanketed with very thick deposits of tephra which has since weathered to rich soils. The land is planted with vines, vegetables, orange and lemon trees, herbs, and flowers. The reason why volcanic rocks make some of the best soils on earth is because they not only have a wide variety of common elements of the rock and but also these elements are readily chemically separated into elemental components. This fertile soil is a result of the breakdown of various minerals – such as olivine, pyroxene, amphibole, and feldspar (the essential ingredients of volcanic ash and lava) which releases iron, magnesium, potassium, and so on, to the soil.

Search terms: volcano and agriculture; tephra soils; Vesuvius

---

6.4.1 The Natural Hazard

As noted in the previous chapter, scientists in the mid-twentieth century embraced the theoretical model of plate tectonics, which was largely derived from the concept of continental drift. Plate tectonics explained why some sections of the earth's crust appeared to be moving towards, away and/or alongside each other and thus causing earthquakes to occur but also as an explanation for the prevalence of volcanoes. On Earth, there are a number of tectonic plates (see Figure 5.1 in the previous chapter) and it was observed that where convergent plate boundaries occurred, there was a greater frequency of volcanoes (see Figure 6.4); a prime example of this is the 'Pacific ring of fire' that fringes around the Pacific Ocean. Divergent tectonic plates (those pulling away from each other) also generate volcanoes, such as the Mid-Atlantic Ridge that lies beneath the Atlantic Ocean.

6.4.2 Types of Volcanoes

Volcanoes are largely caused by the upwelling of magma (molten rock) towards weak spots on the earth's surface (refer to Box 6.1 for explanation of some key scientific terms used in this chapter). The most common perception of a volcano is of a conical mountain, spewing lava and poisonous gases from a crater at its summit; however, this describes just one of the many types of volcano. The features that lead to the development of volcanoes are much more complicated and their structure and behaviour depends on a number of factors (see Table 6.3 for a summary of the key classifications of volcano). These structural differences are important to consider because they have a bearing upon how violent the eruptions can be and thus what impacts the volcanoes can have.

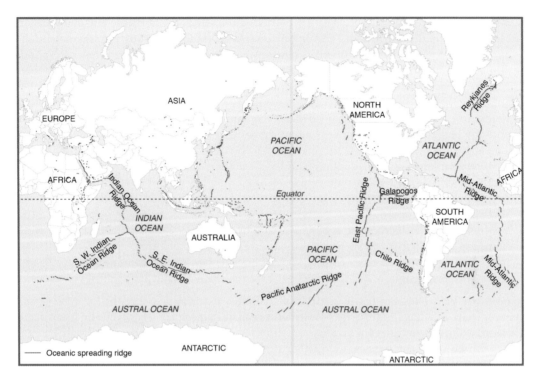

Figure 6.4 Global distribution of major volcanoes (*Source:* WEDC).

6.5 Volcanic Activity

There are several types of volcanic eruptions — during which lava, tephra and assorted gases are expelled from a volcanic vent or fissure. These are often named after famous volcanoes where that type of behaviour has been observed (see Figure 6.5). Some volcanoes may exhibit only one characteristic type of eruption during a period of activity, whereas others may display an entire sequence of types all in one eruptive series.

Key factors:

Main causes: Volcanoes are largely caused by the upwelling of magma (molten rock) towards weak spots on the earth's surface. Typically (but not always) volcanoes occur along tectonic fault lines, such as the convergent plates that circle the Pacific Ocean. Volcanoes are influenced by proximately to, and types of, tectonic plate boundaries and the geological composition of the ground.

Characteristics: As mentioned earlier, the characteristics of volcanoes have a number of influencing factors. Of key concern for this chapter are the characteristics of how the hazard manifests itself, notably when an eruption takes place. Volcanic eruptions are not all the same and are typically categorised into the follow types (also see Figure 6.5), with the most destructive types at the top of the list:

- *Plinian* - Plinian eruptions, named after the eruptions of Mount Vesuvius, are marked by columns of gas and volcanic ash extending high into the stratosphere, the second layer of Earth's atmosphere. The key characteristics are the ejection of large amounts of pumice and very powerful continuous gas blast eruptions.

- *Peléan* – Named after Mount Pelée, a semi-active volcano at the northern end of the Island of Martinique in the Caribbean. The initial phases of these eruptions are characterized by pyroclastic flows. The tephra deposits have lower volume and range than the corresponding Plinian and Vulcanian eruptions. The viscous magma then forms a steep-sided dome or volcanic spine in the volcano's vent. The dome may later collapse, resulting in flows of ash and hot blocks.
- *Vulcanian* – Named after Vulcano, a small volcanic island in the Tyrrhenian Sea, these eruptions are characterised by a dense cloud of ash-laden gas exploding from the crater and rising high above the peak.
- *Strombolian* - Strombolian eruptions consist of ejections of incandescent cinder, lapilli and lava bombs to altitudes of tens to a few hundreds of metres. The eruptions are small to medium in volume, with sporadic violence (see Figure 6.6). They are named after the Italian volcano Stromboli.
- *Icelandic* - The Icelandic type is characterized by effusions of molten basaltic lava that flow from long, parallel fissures. Such outpourings often build lava plateaus.
- *Hawaiian* – These are a type of volcanic eruption where lava flows from the vent in a relatively gentle, low level eruption; it is so named because it is characteristic of Hawaiian volcanoes.

Key impacts: Volcanoes can result in extensive damages that can take months (or even years) to recover from. Buildings and critical infrastructure can be damaged or destroyed and it can become extremely difficult for affected areas to function for a prolonged period; accordingly economic losses can be vast. The impacts are further exacerbated when the volcanic activity generates a

---

Box 6.1

Some Common Volcano Terms Explained

Caldera is a large, usually circular depression at the summit of a volcano formed when magma is withdrawn or erupted from a shallow underground magma reservoir.

Lahar is an Indonesian word for a rapidly flowing mixture of rock debris and water that originates on the slopes of a volcano.

Lapilli are rock fragments between 2 and 64 mm (0.08-2.5 in) in diameter that were ejected from a volcano during an explosive eruption.

Magma/Lava is molten or partially molten rock beneath the Earth's surface. When magma erupts onto the surface, it is called lava.

Phreatic eruptions are steam-driven explosions that occur when water beneath the ground or on the surface is heated by magma, lava, hot rocks, or new volcanic deposits

Phreatomagmatic eruptions are eruptions that arise from interactions between water and magma. They are driven from thermal contraction (as opposed to magmatic eruptions, which are driven by thermal expansion) of magma when it comes in contact with water.

Pyroclastic flow is an avalanche of hot (up to 500° C) ash, pumice, rock fragments, and volcanic gas that rushes down the side of a volcano as fast as 100 km/hour (62 miles/hour).

Scoria is a bubbly glassy lava rock of basaltic to andesitic composition ejected from a vent during explosive eruption. The bubbly nature of scoria is due to the escape of volcanic gases during eruption.

Tephra is generic term for fragments of volcanic rock and lava that are blasted into the air by explosions or carried upward by hot gases in eruption columns or lava fountains.

tsunami (see Box 5.1 in the previous chapter) or effuses significant amounts of debris into the atmosphere resulting in disruption to air travel or even short-term changes to weather patterns (see Case Study 6.3 about the long-term impacts of the eruption of Mount Tambora in 1815).

Table 6.3 Summary of Different Types of Volcanoes and Their Key Features.

Volcano type	Typical features (and examples)
Stratovolcanoes (composite volcanoes)	Stratovolcanoes are also known as composite volcanoes because they are created from multiple structures during different kinds of eruptions. Strato/composite volcanoes are made of cinders, ash that pile up on top of each other, lava flows on top of the ash, where it cools and hardens, and then the process repeats. Classic examples include Mt. Fuji in Japan, Mayon Volcano in the Philippines, and Vesuvius and Stromboli in Italy.
Lava domes	Lava domes are built by slow eruptions of highly viscous (thick) lava. They are sometimes formed within the crater of a previous volcanic eruption, but can also form independently. Like stratovolcanoes, they can produce violent, explosive eruptions, but their lava generally does not flow far from the originating vent. Chaitén in Chile and Mount St. Helens in the USA are good examples.
Shield volcanoes	These volcanoes are named due to their broad, shield-like profiles. They are formed by the eruption of low-viscosity (thin) lava that can flow a great distance from a vent. They generally do not explode catastrophically. The Hawaiian Islands and the Galápagos Islands have examples of these types of volcanoes.
Supervolcanoes	A supervolcano usually has a large caldera and can produce devastation on an enormous, sometimes continental, scale. Such volcanoes are able to severely cool global temperatures for many years after the eruption due to the huge volumes of sulphur and ash released into the atmosphere. They are the most dangerous type of volcano. Examples include Yellowstone Caldera in the Yellowstone National Park and Valles Caldera in New Mexico (both western United States); Lake Taupo in New Zealand; Lake Toba in Sumatra, Indonesia; and Ngorongoro Crater in Tanzania.
Cinder cones	These result from eruptions of mostly small pieces of scoria and pyroclastics that build up around the vent. These can be relatively short-lived eruptions that produce a cone-shaped hill perhaps 30 to 400 meters high. Most cinder cones erupt only once. Cinder cones may form as flank vents on larger volcanoes, or occur on their own. Cerro Negro in Nicaragua, Parícutin in Mexico and Sunset Crater in Arizona are examples of cinder cones.
Fissure vents	These are flat, linear fractures through which lava emerges. Examples include Lanzarote in the Canary Islands in Spain and Holuhraun in Iceland
Submarine volcanoes	Submarine volcanoes are common features of the ocean floor. In shallow water, active volcanoes disclose their presence by blasting steam and rocky debris high above the ocean's surface. In the ocean's deep, the tremendous weight of the water above prevents the explosive release of steam and gases.
Subglacial volcanoes	Subglacial volcanoes develop underneath icecaps. They are made up of flat lava which flows at the top of extensive pillow lavas. When the icecap melts, the lava on top collapses, leaving a flat-topped mountain.
Mud volcanoes	Mud volcanoes or mud domes are formations created by geo-excreted liquids and gases, although there are several processes which may cause such activity. The largest structures are 10 kilometres in diameter and reach 700 meters high.

Figure 6.5 The main types of volcanic eruptions (*Source:* WEDC).

Figure 6.6 Sakurajima Volcano, Vulcanian Eruption, Japan in 2010 (*Source:* Reproduced with permission of Richard Roscoe).

CASE STUDY 6.3

Far-Reaching Impacts of the Mount Tambora Eruption in 1815

In 2009 the findings of a study on the impacts of the 1815 Tambora volcanic eruption were published in the *Journal of Climatology*. The researchers studied the impact of this atmospheric phenomenon on agricultural production in the Iberian Peninsula in 1816 and 1817, and evaluated and compared the climate in the area with that in 1871-1900, before it started to be affected by climate change.

The Tambora volcano in Indonesia erupted in April 1815, but North America and Europe did not notice its effects until months later. In 1816, known as "the year without a summer", gases, ashes and dust arrived over the Iberian Peninsular and reached the stratosphere, where they remained long enough to create "an enormous sun filter". The year 1816 was characterised by great anomalies, especially in the summer, which was much colder and wetter than usual. In Madrid, temperatures were below 15°C in July and August, and that autumn the Catalan peaks of Montserrat and Montseny were covered with snow and the Llobregat river froze over.

The eruption of the Tambora volcano was probably "the greatest recorded eruption in historical times" according to the researchers. This is demonstrated by its explosivity index (a measurement of the size of the eruption), which, at 7, "was greater than any other more recent eruption, including that of Mt. Pinatubo in the Philippines". The consequences of the eruption were noticed not only on the climate, but above all on agriculture. "The low temperatures meant that many crops did not ripen, or if they did their yield was very little and very late," says García Herrera.

> The decade from 1811 to 1820 was marked by serious socio-economic impacts resulting from this poor agricultural production, with malnutrition and the increase of epidemics in Europe and Mediterranean countries. Low temperatures, freezing temperatures in spring and heavy precipitation between 1816 and 1817 affected the growth of many crops very badly. The cold and wet summer led to reduced harvests due to fruits being of poor quality, as well as vines and cereals ripening very slowly.
>
> **Reference:** Trigo, R. M., Vaquero, J. M., Alcoforado, M.-J., Barriendos, M., Taborda, J., García-Herrera, R. and Luterbacher, J. (2009), 'Iberia in 1816, the year without a summer'. *Int. J. Climatol.*, 29: 99–115

The Volcanic Explosivity Index (VEI) was devised by Chris Newhall of the United States Geological Survey and Stephen Self at the University of Hawaii in 1982 to provide a relative measure of the explosiveness of volcanic eruptions (see Table 6.4). Volcanic eruptions are one of Earth's most dramatic and violent agents of change. Not only can powerful explosive eruptions drastically alter land and water for tens of miles around a volcano, but tiny liquid droplets of sulphuric acid erupted into the stratosphere can temporarily change earth's climate. Eruptions often force people living near volcanoes to abandon their land and homes, sometimes forever. Those living farther away are likely to avoid complete destruction, but their cities and towns, crops, industrial plants, transportation systems, and electrical grids can still be damaged by tephra, ash, lahars, and flooding (Figure 6.7).

Location: Scientists and relevant emergency planners/services have a very good idea of where volcanic eruptions can occur based upon the sometimes very obvious physical presence of volcanic mounds/domes and records of previous volcanic activity. However, knowing the possible location of volcanoes is one thing, knowing if (see Box 6.2) or when they may erupt is another more complex matter.

Advance warnings:

A broad range of scientific research, as well as observations from local populations, have identified a number of possible precursors to volcanic eruptions. At the moment this is somewhat an inexact science but it is nonetheless worth acknowledging that such precursors may include one or more of the following activities:

a) seismicity under the volcano (see Case Study **6.4** about Japan),
b) uplift of the land over the volcano (this happened at Mount St. Helens prior to the 1980 eruption),
c) increasing emissions of volcanic gases such as carbon dioxide or sulphur dioxide (see Case Study **6.5** for the findings of some recent research),
d) increase in stream/spring water temperatures near the volcano,
e) increase in the temperature of the land at the volcano.

It is possible to use scientific monitoring equipment to keep a check on some of the aforementioned precursors and this should be undertaken in a way that builds upon the (often perceived to be unscientific) knowledge and observations of local populations. For instance, participatory research conducted on Ambae Island in Vanuatu by Cronin *et al.* (2004) found that the local villagers identified precursors such as pervasive gas smells, the death of trees around Lake Vui, unusually active bubbling and other lake disturbances, lake water colour changes (to brown or grey), rumbling and booming from the crater, and the rapid rotting of taro roots in the ground.

Table 6.4 The Volcanic Explosivity Index (VEI).

VEI	Ejecta volume(bulk)	Classification	Description	Plume	Frequency	Tropospheric injection	Stratospheric injection	Examples
0	$<10,000 m^3$	Hawaiian	Effusive	<100m	constant	negligible	none	Kilauea, Piton de la Fournaise, Erebus
1	$>10,000 m^3$	Hawaiian / Strombolian	Gentle	100m–1 km	daily	minor	none	Nyiragongo (2002), Raoul Island (2006), Stromboli Island - (continuous since Roman times to present)
2	$>1,000,000 m^3$	Strombolian / Vulcanian/ Hawaiian	Explosive	1–5 km	weekly	moderate	none	Unzen (1792), Cumbre Vieja (1949), Galeras (1993), Sinabung (2010)
3	$>10,000,000 m^3$	Vulcanian / Peléan/Sub-Plinian/Hawaiian	Catastrophic	3–15 km	few months	substantial	possible	Nevado del Ruiz (1985), Soufrière Hills (1995), Nabro (2011)
4	$>0.1 km^3$	Peléan / Plinian/Sub-Plinian	Cataclysmic	>10 km (Plinian or sub-Plinian)	≥1 yr	substantial	definite	Mayon (1814), Pelée (1902), Galunggung (1982), Eyjafjallajökull (2010)
5	$>1 km^3$	Peléan/Plinian	Paroxysmic	>10 km (Plinian)	≥10 yrs	substantial	significant	Vesuvius (79), Fuji (1707), Mount Tarawera (1886), St. Helens (1980), Puyehue (2011)

(Continued)

Table 6.4 (Continued)

VEI	Ejecta volume(bulk)	Classification	Description	Plume	Frequency	Tropospheric injection	Stratospheric injection	Examples
6	>10 km^3	Plinian / Ultra-Plinian/ Ignimbrite	Colossal	>20 km	≥ 100 yrs	substantial	substantial	Veniaminof (c. 1750 BC), Huaynaputina (1600), Krakatoa (1883), Novarupta (1912), Pinatubo (1991), Laacher See (c. 12,900 BC)
7	>100 km^3	Ultra-Plinian/Plinian/ Ignimbrite	Super-colossal	>20 km	$\geq 1,000$ yrs	substantial	substantial	Mazama (c. 5600 BC), Thera (c. 1620 BC), Taupo (180), Samalas (Mount Rinjani) (1257), Tambora (1815)
8	$>1,000$ km^3	Ignimbrite/Plinian/ Ultra-Plinian	Mega-colossal	>20 km	$\geq 10,000$ yrs	vast	vast	La Garita Caldera (26.3 Ma), Yellowstone (640,000 BC), Toba (74,000 BC), Taupo (24,500 BC)

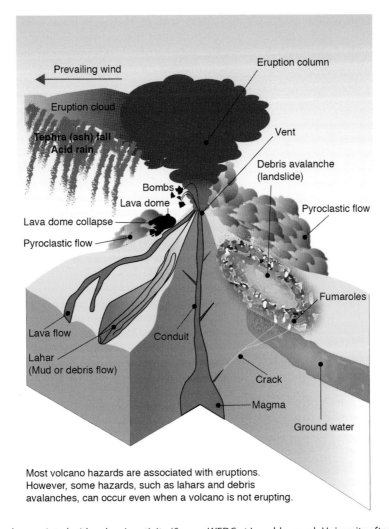

Most volcano hazards are associated with eruptions.
However, some hazards, such as lahars and debris
avalanches, can occur even when a volcano is not erupting.

Figure 6.7 Hazards associated with volcanic activity (*Source:* WEDC at Loughborough University after USGS).

6.6 Risk Management

6.6.1 Historical Approaches

As highlighted in early chapters, in the past societies have tended to be aware of their natural environment and planned their settlements and built their homes and infrastructure in a way that minimises any disruptions from natural hazards. Volcanoes have historically attracted human settlements, largely due to the nutrient rich soils that have benefitted agricultural production but also to the advantages of local geothermal activity.

Historical records have been testament to ancient civilisations suffering from the impacts of volcanoes, such as the Minoan civilisation on Thera (1500 BC in modern day Greece) and the Roman cities of Pompeii and Herculaneum (AD 79 in modern day Italy). There are also recent examples of humankind's sometimes fragile relationship with volcanoes, for instance, the Nevado del Ruiz eruption in Colombia (1985) and the Soufrière Hills eruptions (1995 in Monserrat). Despite this long-term, and sometimes well documented history there is little evidence that past societies have actually done much to reduce the impacts of volcanic eruptions, but this is arguably with good reason.

Box 6.2

Is That Volcano Active, Dormant or Extinct?

Scientists, particularly geologists, classify most volcanoes as either 'active' or 'dormant'. However, classification of volcanoes is complicated because there's no comprehensive way of knowing whether a volcano is ever going to be active again.

An active volcano is one that's currently in a state of 'regular' eruptions. Maybe it has erupted recently and has also had an event in the last few decades. A dormant volcano is one that is capable of erupting, and will probably erupt again in the future, but it hasn't had an eruption for a 'very long time'. However, you will note that some of these terms (i.e., 'regular' and 'very long time') are quite vague. The problem is that the lifespan of a volcano can last for thousands of years, or it can go on for millions of years, with regular eruptions. Many of Earth's volcanoes have had dozens of eruptions in the last few thousand years, but they've been quiet for recorded history, and have large populations built up around their base. The Smithsonian Global Volcanism Program defines a volcano as active if it has had an eruption within the last 10,000 years or so.

Therefore, a dormant volcano is actually part of the active volcano classification meaning that a volcano has not currently erupted. When a volcano becomes cut off from its magma supply that is when it finally stops erupting and becomes an extinct volcano. For instance, Huascarán is Peru's highest mountain at 22,205 feet (6,768 meters) high and is considered to be an extinct volcano. Not all extinct volcanoes have to be so tall, with Edinburgh Castle being built upon an extinct volcano in the Scottish capital city.

CASE STUDY 6.4

Monitoring of Volcanic Activity in Japan

There are 110 active volcanoes in Japan (Figure 6.8); on average, a total of 15 volcanic events (including eruptions) occur every year, some of which seriously hinder human life. To continuously monitor this volcanic activity, the Japan Meteorological Agency (JMA) deploys seismographs and related observation instruments in the vicinity of 47 volcanoes that are most significantly active. Mobile observation teams are also sent to other volcanoes for regular patrols. When volcanic anomalies are detected, the Agency steps up its monitoring/observation activities and publishes volcanic information and regular bulletins.

In order to detect unusual volcanic phenomena and issue volcanic information appropriately, JMA operates Volcanic Observations and Information Centers at JMA Headquarters and at the Sapporo, Sendai and Fukuoka Regional Headquarters of JMA, which integrate various types of observation data and monitor volcanic activity in their areas of responsibility. JMA began issuing Volcanic Warnings and Volcanic Forecasts for each active volcano in Japan in December 2007 to mitigate damage from volcanic activity. Volcanic Warnings are issued in relation to expected volcanic disasters, and specify the municipalities where people need to take action. Volcanic Forecasts are issued for less active volcanoes or those that become so.

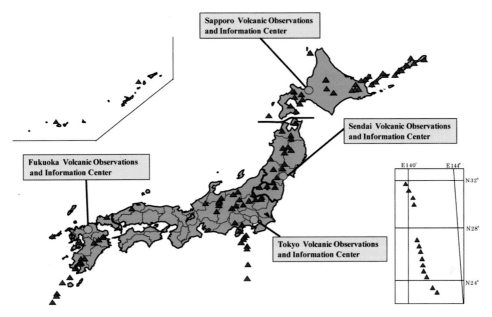

Figure 6.8 Active volcanoes in Japan (*Source:* Japan Meteorological Agency).

CASE STUDY 6.5

New 'Early Warning Sign' for Volcanic Eruptions

Recent research has found that volcanoes are primed to erupt on a timescale of days to months rather than years. The study raises hopes of finding a more accurate way to tell when a volcano is about to explode.

A team of geologists has reported that gases emitted from the mouth of a volcano could provide clues to when an eruption is imminent. They studied the dormant Campi Flegrei volcano that has been dubbed Europe's super volcano and which last erupted in 1538. Crystals of a mineral from the ancient explosion show the molten rock was 'primed' – or filled with bubbles of gas – only very shortly before erupting.

Mike Stock, lead researcher on the study, said volcanoes can be 'primed' on short time scales, which will help scientists interpret the signals expected to appear before such an event. "When the magma forms bubbles, the composition of gas at the surface should change, potentially providing an early warning sign," he said. More work is planned to try to confirm the findings. The research was carried out by the University of Oxford with Durham University and the National Institute of Geophysics and Volcanology, Rome.

Reference: Stock, M.J., Humphreys, M.C., Smith, V.C., Isaia, R. and Pyle, D.M., (2016), 'Late-stage volatile saturation as a potential trigger for explosive volcanic eruptions', *Nature Geoscience*.

Compared to many of the other hazards discussed in this book, there is acknowledgement that there not anything that can be done to prevent a volcanic eruption. It is also recognised that there is little that can be done to limit the impacts or protect communities from the impacts of volcanic activity, particularly the highly damaging lava flows, pyroclastic flows, boulder/ash falls and lahars. To further complicate matter, long intervals between eruptions can contribute to public ignorance of, or complacency about, the potential threat. Consequently, most of the disaster risk reduction activities discussed over the next sections of this chapter are primarily related to detection and emergency response/preparedness.

6.7 Risk Management

6.7.1 Hazard Identification

Volcanic hazard identification can be performed in numerous ways. These can be categorised into three principal groups: desk study, site inspection, and detailed investigation. It is typically the engineer (geotechnical engineer or engineering geologist) who will carry out the hazard identification process (these are explained in more detail in Chapter 7 on Landslides, see Table 7.3). The location (and thus ongoing activities) of the world's 500 plus active volcanoes is generally well known so the main uncertainties tend to be associated with when, if ever, they will erupt and the potential magnitude of such eruptions (Table 6.5).

6.7.2 Assessment of the Vulnerabilities

Once an understanding has been obtained of the extent to which volcanic hazards are prevalent, it is then critical to gain an appreciation of what types of sites, communities, buildings, facilities and

Table 6.5 Overview of Typical Volcanic Hazard Identification Methods.

Context	Hazard identification methods
Historical experience	We know where volcanoes can erupt because they tend (but not always) to be quite obvious features on the landscape and previous (recent) incidents might have been recorded and reported by national or local geological survey institutions such as the British Geological Survey (BGS) in the UK, the Geological Survey of Japan (GSJ) or the U.S. Geological Survey (USGS). Of course, historical incidents should be treated with caution because volcanoes can lie dormant for hundreds of years between eruptions.
Geological mapping	Scientific advances in the last few centuries have contributed towards the development of high-resolution mapping techniques and monitoring of volcanic activity. Maps can be generated that highlight areas most susceptible to different types of volcanic hazards, such as pyroclastic flows, lava and lahars (see Figure 6.9). These maps can also be used for estimating potential economic losses and preparing for emergency response. It should be noted that these hazard maps may not to be accurately predict the direction of any ash fall as this will typically be determined by localised and often dynamic weather patterns.
Monitoring and warning	A typical volcano monitoring and warning system can be made up of a system of seismographs, gas emission monitors and alarms that are designed to monitor long-term trends and also to notify local agencies of any substantial seismic activity and/or releases of indicator gases before a potential eruption. Again, this is not really a pure or well test science but developments are being made by the scientific communities.
Prediction	Being able to predict when and where a volcano will erupt could save thousands, if not hundreds of thousands, of lives. Even after decades of study, volcanic prediction remains notoriously difficult, however there has been an increasing interest in the identification or 'precursors' that may act as early indicators of an eruption (also see Case Study **6.5**). Accurate early warnings of a volcanic eruption could provide emergency managers and local communities with valuable time to evacuate at risk areas

services could be affected by these hazards. This involves an assessment of the potential vulnerabilities of specific aspects of such locations that can be categorised (previously detailed in Chapter 2) as shown in Table 6.6.

6.7.3 Determination of the Risk

Effects associated with volcanoes can be divided into primary and secondary. The primary effects are those caused by the actual volcanic eruption. The secondary effects are those triggered by the primary impacts (see Table 6.7 for more examples).

6.7.3.1 Primary Volcanic Hazards

- *Lava flows:* A greater threat to property than human life due to opportunity for evacuation. Lava flows are more dangerous when released very quickly. They usually cause large but localised economic losses.
- *Pyroclastic flows:* These are hot rock fragments, lava particles ash and hot gases. They are linked with subduction zone volcanoes, and the flow moves very quickly from its source. This hazard is sometimes pyroclastic density currents (PDCs)
- *Ash and Tephra fall:* Ash is material below two millimetres in diameter whilst tephra is anything above this. It is usually formed when magma is fragmented by explosions, and can stay in the

Table 6.6 Overview of Typical Volcanic Hazard Identification Methods.

Type of vulnerability	Examples related to volcanic hazards
Physical vulnerability	<ul><li>Located in areas with high topographical relief and high population density.</li><li>River valley below slopes that could become channels for pyroclastic flows, lahars and/or lava flows</li><li>Proximity of critical infrastructure to areas likely to be affected by volcanic hazards</li><li>Location of building/infrastructure in relation to the potential volcanic hazards</li><li>Remoteness of settlement</li></ul>
Social vulnerability	<ul><li>Inadequate awareness of volcanic risk and emergency evacuation procedures (if any)</li><li>Lack of guidance and regulations on how to avoid areas most prone to the impacts of volcanic.</li><li>Inability for some individuals (e.g., young children and the elderly) to evacuate in a timely manner in the event of an eruption.</li><li>Little choice over where to construct a home (i.e., only option to build in an area prone to pyroclastic or lava flows).</li></ul>
Economic vulnerability	<ul><li>Unable to afford even basic monitoring and early warning systems.</li><li>Limited capacity to provide response mechanisms to evacuate, rescue and remediate.</li><li>Lack of economic status leading to little choice over where to construct a home.</li></ul>
Environmental vulnerability	<ul><li>Over farming of poor soils pushing farmers to live nearer to the more fertile soils located near volcanoes</li></ul>
Governance vulnerability	<ul><li>Lack of guidance and regulations on how to take into account volcanic risk in the design and construction of built environment infrastructure.</li><li>Lack of understanding of how to respond to volcanic events.</li><li>Inadequate infrastructure and services in place to respond to volcanic eruptions.</li><li>Inadequate emergency management strategies.</li></ul>

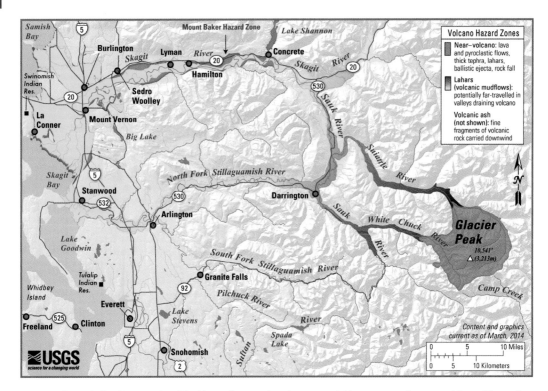

Figure 6.9 Crater Lake, Oregon, simplified hazards map showing potential impact area for ground-based hazards such as lava and lahars during a volcanic event (*Source:* USGS).

atmosphere causing global variations in weather patterns. Ash fall does not cause many deaths but can lead to breathing difficulties. The Eyjafjallajökull volcano (Iceland) in 2010 was particularly problematic because the volcanic ash plume disrupted air travel in northern Europe for several weeks.

- **Volcanic gases:** Gases emitted from volcanoes include, water vapour, carbon dioxide, sulphur dioxide, hydrogen sulphide, helium and carbon monoxide. They rarely cause death but can be problematic as they tend to be heavier than air.

6.7.3.2 Secondary Volcanic Hazards

- **Lahars:** Mud flows of volcanic material due to ash and debris mixing with water. On steep slopes the speeds of these flows can reach 22 metres per second.
- **Volcanic landslides:** These are falls of rock and loose volcanic material, which are driven by gravity.
- **Tsunamis:** These can be quite rare events, but the 1883 eruption of Krakatoa, did create a 30m (98 feet) high tsunami that have wide reaching impacts.

Wilson *et al.* (2014), provide an extensive review of the main vulnerabilities for critical infrastructure sectors from the impacts of volcanic hazards. As part of the review the authors provide information about whether the impacts from four specific hazards (tephra/ash fall, pyroclastic flow, lava flow and lahar) can be mitigated by site exclusion (avoidance), physical design of infrastructure or response and operational planning. Table 6.8 gives some examples from Wilson *et al.* (2014), with particular focus on vulnerabilities to transport networks and buildings. A clear message from this

Table 6.7 Generic Examples of the Primary and Secondary Effects of Volcanos.

	Primary effects	Secondary effects
Physical	*Lava/pyroclastic flows, volcanic gases and ash falls leading to damage to...* • Critical infrastructure • Houses • Offices/shops • Schools • Vehicles • Equipment • Agriculture and livestock • Water courses	*Including fires, tsunamis, landslides, resulting in....* • Homelessness/displacement • Disrupted essential services • Reduced access for heavy plant/vehicles • Contaminated water channels/sewers • Polluted land and water sources • Disrupted education provision • Loss of crops (in the short term)
Wellbeing/ health	*Polluted air and Deaths or injuries to...* • People • Livestock • Fish stocks	*Leading to...* • Post-traumatic stress disorder • General fear of safety due to aftershocks • Unhygienic conditions • Destroyed land/homes/villages • Strains upon social support services • Respiratory conditions (short and long term)
Economic	*Immediate...* • Repair/replacement costs • Loss of business • Loss of tourism	*Longer term...* • Rehabilitation costs • Insurance costs (and increased premiums) • Reduced commerce • Loss of crops (but potential long-term increases in soil fertility) • Loss of tourist revenues • Reduced land values

table is that for the impacts of pyroclastic and lava flows and lahars, infrastructure development should be avoided in hazard prone sites because damage from these types of hazards cannot be mitigated.

6.8 Identification and Prioritisation of Risk Reduction Options

As highlighted in earlier chapters, there can be a number of ways to reduce the risk of a hazard. However, in the case of volcanoes there are very limited options available when compared to dealing with other hazards such as floods (see Chapter 3). As outlined in Chapter 2, the best approach to considering risk reduction measures is the adoption of the five interrelated stages as shown in Table 6.9, which summarises the extent to which each of the risk reduction options can be utilised for volcanoes (excluding the impacts of tsunamis, which are covered in Chapter 5). A more detailed, if not rather limited, list of specific risk reduction examples is provided for dealing with volcanic risk in Table 6.10.

It is very important to acknowledge at this stage that the infrequency of volcanic eruptions can induce poor awareness of the hazard and thus low levels of community or even institutional preparedness. Therefore, monitoring of volcanic activity and timely warning mechanisms are important, but these also need to be supported with local capacity building activities aimed at raising public awareness of the hazards as well as developing and testing evacuation procedures.

6.8.1 Inherent Safety and Prevention

The only way that this could be achieved would be to avoid building (or indeed demolished existing developments) on land that is prone to volcanic activity. This approach may be possible for undeveloped land/regions but in the twenty-first century we are dealing with a legacy of human habitations that have been built in volcanically active areas due to the agricultural benefits such areas have provided.

Table 6.8 Summary of the Main Vulnerabilities and Considerations for Selective Infrastructure From Key Volcanic Hazards (after Wilson *et al.* 2014).

	Volcanic Hazards			
Infrastructure type	**Tephra/ash fall**	**Pyroclastic flows**	**Lava flows**	**Lahars**
Transportation networks	*Vulnerability*: Reduced visibility and traction, covering of road and runway markings, abrasion and corrosion damage to vehicles, jamming of rail switches, and disruption to airspace.	*Vulnerability*: Burial of roads, rail networks and airport runways, increased sedimentation into harbours, erosive damage and destruction of bridges, extensive damage to vehicles.	*Vulnerability*: Burial of roads, rail networks and airport runways.	*Vulnerability*: Burial of roads, rail networks and airport runways, increased sedimentation into harbours, erosive damage and destruction of bridges, extensive damage to vehicles.
	Site exclusion: No	*Site exclusion*: Yes — where possible all routes should be located away from valleys and known flow paths.	*Site exclusion*: Yes — where possible all routes should be located away from known flow paths.	*Site exclusion*: Yes — where possible all routes should be located away from valleys and known flow paths.
	Design: Strengthen buildings (airports, train stations) and increase roof pitch to minimise tephra load damage.	*Design*: Raise bridge decks over valleys and strengthen piers and abutments.	*Design*: Construction of embankments around critical parts of the network.	*Design*: Automated barriers to close road and rail routes when lahars occur. Raise bridge decks over valleys and strengthen piers and abutments.
	Contingency planning: Tephra clean-up operations and methods. Road, rail and airport closure protocols. Established tephra avoidance guidelines for aircraft.	*Contingency planning*: Identify alternate routes if primary routes are damaged. Anticipate the need for temporary bridges. Clean-up operations and methods.	*Contingency planning*: Identify alternate routes if primary routes are damaged. Anticipate the need for temporary bridges.	*Contingency planning*: Use of early warning systems. Identify alternate routes if primary routes are damaged. Anticipate the need for temporary bridges.

Table 6.8 (Continued)

Infrastructure type	Volcanic Hazards			
	Tephra/ash fall	Pyroclastic flows	Lava flows	Lahars
Buildings	*Vulnerability*: Blocked and/or damaged gutters, tephra ingress, corrosion of metal surfaces, and structural damage to roof.	*Vulnerability*: Damage to windows and doors, structural damage to whole building, inundation and burial, ignition of fires.	*Vulnerability*: Structural damage to whole building, burial, ignition of fires.	*Vulnerability*: Inundation and burial, structural damage to walls, float building off foundations.
	Site exclusion: No	*Site exclusion*: Yes — where possible all buildings should be located away from valleys and known flow paths.	*Site exclusion*: Yes — where possible all buildings should be located away from known flow paths.	*Site exclusion*: Yes — where possible all buildings should be located away from valleys and known flow paths.
	Design: Strengthen roofs, increasing roof pitch to reduce static load.	*Design*: Strengthen walls and avoid having them perpendicular to flow path to reduce dynamic load. Use of shutters on openings to prevent ingress.	*Design*: Strengthen building walls. Use of non-flammable materials.	*Design*: Strengthen walls and avoid having them perpendicular to flow path to reduce dynamic load. Fix buildings to foundations.
	Contingency planning: Sealing of building to prevent tephra ingress. Removing tephra from roof to prevent collapse.	*Contingency planning*: Evacuation planning and implementation.	*Contingency planning*: Use of early warning systems. Evacuation planning and implementation.	*Contingency planning*: Use of early warning systems. Evacuation planning and implementation.

Notes:
a) A 'yes' for site exclusion indicates that infrastructure development should be avoided at a particular site as damage from a hazard cannot be mitigated.
b) Design considerations include altering the design of components and infrastructure sectors to lower their vulnerability to disruption and damage from volcanic hazards (e.g., strengthen building roof) and the design of site protection measures for flow hazards (e.g., construction of diversion barriers).
c) Contingency planning involves making decisions and plans in advance about the management and response to volcanic eruptions to minimise impact severity and decrease recovery time (e.g., evacuation plans, clean-up plans and availability of resources).

6.8.2 Detection of Hazard

It may be possible to detect volcanoes but at the moment the science behind the detection systems provides us with limited benefits. Early warning systems can provide minutes, hours or potentially even days of advance warning that would be critical to allow people and critical infrastructure systems to take actions to protect life mainly through evacuation of the area (refer to Case Studies 6.4 and 6.5 for some examples).

Table 6.9 Summary of the Viability of Risk Reduction Options for Addressing Volcanic Risks.

Inherent safety	Type of risk reduction option for volcanic risks				
	Prevention	Detection	Control	Mitigation	Emergency response
N	N	#	N	#	#

Please note:
'Y' – indicates that there are possibly a range of useful risk options available
'#' – indicates that some risk reduction options can be used but they are likely to be of only limited effectiveness
'N' – indicates that other than relocating the built asset there is little that can be done to reduce this hazard/threat

6.8.3 Control of the Hazard

If a volcano erupts, there is little, arguably nothing that can be done to reduce the amount of lava, pyroclastic flows or ash fall that will be generated. Therefore, it is not currently possible to control the hazard.

6.8.4 Mitigation of Hazard

Volcanic risk can best be reduced through sound local planning, that sets out to avoid areas prone to the main volcanic hazards. This would include avoiding construction of buildings and critical infrastructure in river valleys and gullies running down from the volcanic cone. This may also include avoidance of development in low lying areas prone to lava flows. Large walls can be built to mitigate the impact of lahars but the success of these types of expensive and intrusive structures has thus far been rather limited (see Figure 6.10 and 6.11 for examples). Clearly, inspections should be conducted to check for building safety (i.e., government buildings, hospitals, schools and other critical infrastructure) to ensure the structures are designed to cope with seismic activity and potentially heavy ash falls. It may be useful to seal some key buildings (particularly healthcare facilities) to prevent tephra/ash ingress. If tephra/ash fall is significant then it is important that such debris is removed from roofs to prevent collapse (see Figure 6.12).

6.8.5 Emergency Response

Preparedness planning; as it is widely acknowledged how difficult it is to provide sufficiently long warnings for volcanic eruptions, thus preparedness planning plays an especially important role in enabling affected communities to cope with and recover from damaging volcanic activity. The New Zealand Government's 'Get Ready Get Thru' civil defence campaign recommends that the general public and businesses undertake the following activities when preparing for volcanoes:

- Find out about the volcanic risk in the community. Ask the local council about emergency plans and how they will warn you of a volcanic eruption.
- Practice the evacuation plans with members of the household/business.
- Develop a Household Emergency Plan. Assemble and maintain Emergency Survival Items for home and business as well as a portable getaway kit.
- Include pets and livestock in these emergency plans.

Table 6.10 Indicative Examples of Risk Reduction Options for Addressing Volcanic Risk.

Risk reduction option	Examples
1) Inherent safety – eliminate the possibility of hazards occurring	Not applicable, we cannot remove an active volcano.
2) Prevention – reduce the likelihood of hazards	Not applicable, we cannot prevent volcanoes from erupting.
3) Detection – measures for early warning of hazards	Potentially useful, although as explained in earlier sections of this chapter, the reliability of some identified precursors to volcanic eruption is yet to be validated. A typical volcano monitoring and warning system can be made up of a system of seismographs, gas emission monitors and alarms that are designed to monitor long-term trends and also to notify local agencies of any substantial seismic activity and/or releases of indicator gases before a potential eruption. Accurate early warnings of a volcanic eruption could provide emergency managers and local communities with valuable time to evacuate at risk areas
4) Control – limiting the size of the hazards	Not applicable.
5) Mitigation and adaptation – protection from the effects of hazards	Volcanic risk can best be reduced through local planning, that is, avoiding areas prone to the main volcanic hazards (i.e., avoiding slopes and gullies running down from the volcanic cone or low lying areas prone to lava flows). Large walls can be built to mitigate the impact of lahars but the success of these types of expensive and intrusive structures has been limited. Inspections should be conducted of building safety (i.e., government buildings, hospitals, schools and other critical infrastructure). Sealing of buildings to prevent tephra ingress. Strengthen walls and avoid having them perpendicular to flow path to reduce dynamic load. Fix buildings to foundations.
6) Emergency response – planning for evacuation emergency access	Draw up detailed vulnerability maps. This could include developing an inventory of public and commercial buildings assets that may be particularly vulnerable to volcanic hazards (i.e. routes of potential lava/pyroclastic flows). Identify safe evacuation and emergency access routes. This should go hand in hand with developing an outreach program about evacuation plans and routes in homes, schools, and businesses; and regularly test those plans Ensure that the emergency services stations and facilities are located in areas away from predicted routes of lava/pyroclastic flows and potential lahars.

Draw up detailed vulnerability maps*.* This activity could include local authorities, communities or businesses developing an inventory of public and commercial buildings assets that may be particularly vulnerable to volcanic hazards. A key part of this approach is ensuring that the emergency services stations and facilities as well as (ideally when feasible) the transport routes to and from those facilities are located in area less likely to be affected by surface volcanic hazards such as lava and pyroclastic flows.

Identify safe evacuation and emergency access routes*;* can be of particular relevance when considering the risk of pyroclastic flows and lahars where a warning of even a few minutes can save lives. Evacuation routes can be horizontal where the local topography enables evacuation to nearby hills of 10 meters or higher. Of course, these evacuation procedures are highly reliant on effective monitoring and warning systems.

(a)

(b)

Figure 6.10 Types of lahar diversion structures: (a) Engineered channel reach in small river draining Sakurajima volcano in southern Japan, where channel is revetted with reinforced concrete and engineered to be as steep, narrow, and smooth as possible, in order to divert lahars away from a developed area. (b) Training dike revetted with steel sheet piles on the lower flank of Usu volcano, Japan and designed to deflect lahars away from buildings and other infrastructure (*Source:* Reproduced with permission of Thomas C Pierson).

(a)

(b)

Figure 6.11 Examples of impermeable lahar flow- and erosion-control structures. (a) Series of sheet-pile check dams with masonry aprons at Mount Usu, Japan. (b) Dam of rock-filled steel cribs at Mount Ontake, Japan (*Source:* Reproduced with permission of Thomas C Pierson).

Figure 6.12 An example of urban design used to mitigate the effects of ash fall in Kagoshima, Japan. The design of the housing includes gutter-free roofs (which have a storm water channel located directly beneath), ash-resistant tiles, heavy-duty rubber window and door seals (to aid air tightness to prevent ash entering houses) and large overhanging roofs over balconies (*Source:* Reproduced with permission of David Johnston).

6.9 Summary

Volcanoes are large-scale natural processes that can pose great dangers to local developments. Unlike most of the other hazards discussed in this book, it is widely agreed that there is actually very little we can do to prevent or reduce volcanic hazards. Volcanos tend to be rather obvious features on the landscape, but the extent to which they are active or dormant is less clear and as a consequence, long periods of volcanic inactivity can result in complacency within the local communities and institutions. This complacency can sometimes lead to human induced activities that result in development becoming exposed to volcanic hazards; these activities include:

- Poor (or unregulated) urban planning (building homes, offices, infrastructure and essential services in locations at risk of volcanic hazards)
- Lack of, or ineffective, emergency preparedness and evacuation procedures.

Key points:

- There is very little that can be done to prevent a volcanic eruption or prevent ash, lava and pyroclastic flows from being released.
- If a volcano erupts, there is not much that can be done to reduce the amount of lava, pyroclastic flows or ash fall that will be generated. Therefore, it is not currently possible to control the hazard.
- Traditional knowledge and scientific monitoring approaches have provided some hope in being able to accurately predict volcanic eruptions. It is these approaches that can be critical in providing the minutes or hours advance warning so that exposed populations can be evacuated to safer areas.
- A key part of this approach is ensuring that the emergency services stations and the transport routes to and from those facilities are located in areas less likely to be affected by surface volcanic hazards such as lava and pyroclastic flows.

Reinforcement

- What 'scale' is used to provide a relative measure of the explosiveness of volcanic eruptions?
- What are the most destructive impacts of volcanic eruptions?
- What are the fundamental differences between active, dormant or extinct volcanoes?

Questions for discussion

- Is it fair to state that disasters caused by volcanoes are actually 'natural disasters'?
- Are there any structural risk reduction measures that can be used to deal with volcanoes?
- What are the best ways of raising awareness of what people should do to prepare for a possible volcanic eruption?

Further Reading

Books:
Chester, D.K., (1994). *Volcanoes and society, E.* Arnold, London
Smith, K., (2013). *Environmental Hazards: Assessing Risk and Reducing Disaster.* (6th Edition), Routledge

Articles:
Becker, J.S.; Saunders, W.S.A.; Robertson, C.M.; Leonard, G.S.; Johnston, D.M., (2010). 'A synthesis of challenges and opportunities for reducing volcanic risk through land use planning in New Zealand', *Australasian Journal of Disaster and Trauma Studies*, 2010–11
Cronin, S.J., Gaylord, D.R., Charley, D., Alloway, B.V., Wallez, S. and Esau, J.W., (2004). 'Participatory methods of incorporating scientific with traditional knowledge for volcanic hazard management on Ambae Island, Vanuatu', *Bulletin of Volcanology*, 66(7), pp. 652–668
Kelman, I. and Mather, T.A., (2008). 'Living with volcanoes: the sustainable livelihoods approach for volcano-related opportunities', *Journal of Volcanology and Geothermal Research*, 172(3), pp. 189–198
Pierson, T.C., Wood, N.J. and Driedger, C.L., (2014). 'Reducing risk from lahar hazards: concepts, case studies, and roles for scientists' *Journal of Applied Volcanology*, 3(1), pp. 1–25

Sparks, R.S.J., Biggs, J. and Neuberg, J.W., (2012). 'Monitoring volcanoes', *Science*, 335(6074), pp. 1310–1311

Stock, M.J., Humphreys, M.C., Smith, V.C., Isaia, R. and Pyle, D.M., (2016). 'Late-stage volatile saturation as a potential trigger for explosive volcanic eruptions', *Nature Geoscience*, 9(3), pp. 249–254

Wilson, G., Wilson, T.M., Deligne, N.I. and Cole, J.W., (2014). 'Volcanic hazard impacts to critical infrastructure: A review', *Journal of Volcanology and Geothermal Research*, 286, pp. 148–182

Useful websites:

https://volcanoes.usgs.gov/index.html
http://www.jma.go.jp/jma/indexe.html
http://www.fema.gov/

7

Landslides
Alister Smith[1]

Thousands of people are killed globally each year by landslides. They damage built environment infrastructure, which costs billions of pounds to repair, they leave thousands of people homeless, and they cause the breakdown of basic services such as water supply and transport. There is an urgent need for greater awareness of the ways to manage landslide risk.

7.1   Learning Objectives

By the end of this chapter you will learn:

- What are the different types of landslides?
- What are the main causes of landslides?
- What are the typical impacts of landslides?
- How to identify the risk of landslides and assess vulnerabilities?
- How to reduce the risk of landslides using structural and non-structural measures?

7.2   What are Landslides?

Landslides are masses of soil and/or rock that detach from a slope and travel in a downward motion due to gravity. Landslides come in various forms and fail as a consequence of many different mechanisms. Their sizes range from a falling boulder to a submarine mass movement consisting of several cubic kilometres of earth. Their velocity can range from millimetres per year to kilometres per hour. Thousands of people are killed globally each year by landslides. They damage built environment infrastructure, which costs billions of pounds to repair (see Thinking Point 7.1), they leave thousands of people homeless, and they cause the breakdown of basic services such as water supply and transport.

---

1 Dr Alister Smith is a Lecturer in Infrastructure in the School of Civil and Building Engineering at Loughborough University. He is a Civil Engineer specialising in Geotechnical Engineering and Intelligent Infrastructure.

Disaster Risk Reduction for the Built Environment, First Edition. Lee Bosher and Ksenia Chmutina.
© 2017 John Wiley & Sons Ltd. Published 2017 by John Wiley & Sons Ltd.

Figure 7.1 Landslide damage to a highway in Peace River, Alberta, Canada (*Source:* Reproduced with permission of Matthew Spriggs).

Thinking Point 7.1

The Cost of Landslide Damage

Fatalities from landslides are rare in many parts of the world, but the cost to maintain and remediate infrastructure and the built environment as a result of landslides is high. Therefore, the greatest risk posed by landslides in much of the world is associated with economic loss as opposed to loss of life.

For example, economic losses that were accumulated due to the Holbeck Hall landslide, Scarborough, UK, which was a coastal landslide that destroyed a hotel but did not result in fatalities, were estimated to be £3.5 million (£2 million from insurance claims and £1.5 million for the emergency protection plan) (Gibson *et al.*, 2013).

Moreover, reactivated landslides that move seasonally each year, in response to rainfall, cause annual expenses over consecutive years of the order of millions of pounds due to structural damage, insurance costs, engineering measures and remediation (these cost estimates relate mostly to direct effects; little information is available on indirect costs associated with disruption to traffic and the local economy) (Gibson *et al.*, 2013).

The cost of emergency repair in response to landslides can be ten times greater than the cost of planned maintenance if conducted prior to collapse (Glendinning *et al.*, 2009). This highlights the growing need for effective real-time monitoring; to detect and communicate slope instability early enough to allow for responsive action to be taken (e.g., evacuation or remediation).

Useful search terms:
Landslide damage; landslide remediation; Holbeck Hall landslide, Scarborough

7.3   Statistics on Landslides

EM-DAT records show that in the 50 years between1964 and 2013 there were 583 significant land-slides globally killing nearly 37 thousand people, affecting nearly 10 million people and causing an estimated US$8.5bn in damages. While landslides may not always grab the mass media's attention (compared to the impacts of earthquakes and floods), Table 7.1 shows that the annual impacts of landslides globally can be very significant.

Table 7.2 lists some of the most devastating landslides that have occurred during the last century. More than 32,000 people were killed by non-seismically induced landslides during the period 2004 to 2010 (Petley, 2012). This number is significantly greater with earthquake-induced landslide fatali-ties. Figure 7.2 shows clusters of the non-seismically induced fatal landslide locations superimposed on a world map. The dominant landslide trigger during this period was rainfall in the northern hemi-sphere monsoon and the majority of fatalities occurred in Asia, along the Himalayan Arc and in China (Petley, 2012). Reasons for such geographical clustering include:

- the availability of topographical relief upon which a landslide can occur;
- the frequency and magnitude of precipitation;
- the population density (i.e., the probability of a landslide interacting with people).

Therefore, areas with high topographical relief that experience intense rainfall and have high popu-lation density are most likely to experience fatal landslides.

Table 7.1 Landslide Averages Per Year Between 1964 and 2013.

Major landslides	12
Deaths	737
Total people affected	189,102
Total economic damages (US$)	171m

Table 7.2 List of 10 of the Most Devastating Landslides in the Last Century.

Year	Location	Description	Death toll
1949	Khait, Tajikstan	Earthquake-induced landslides	28,000
1985	Armero, Colombia	Volcanically induced landslides	23,000
1970	Yungay, Peru	Earthquake-induced landslides	22,000
1999	Vargas, Venezuela	Rainfall-induced landslides	10,000+
2013	Kedamath, India	Rainfall-induced landslides	5,700
1963	Vajont Dam , Monte Toc, Italy	A landslide caused a tsunami (seiche) in the impounded reservoir, which overtopped the dam	1,918
2010	Gansu, China	Rainfall-induced landslide	1,700
2011	Rio de Janeiro, Brazil	Rainfall-induced landslides	900+
1966	Aberfan, Wales	Rainfall-induced landslide	144
2014	Oso, Washington State, USA	Rainfall-induced landslide	44

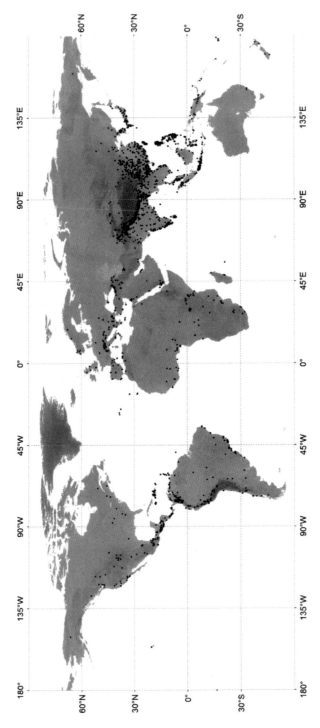

Figure 7.2 Spatial distribution of fatal non-seismically induced landslides (*Source:* Petley 2012. Reproduced with permission of Geological Society of America).

The global population has increased from 1 billion to more than 7 billion over the last 100 years. The majority of this growth over the most recent decades has occurred in low- and middle-income countries. This has driven the expansion of settlements and infrastructure into land increasingly exposed to natural hazards. Coupled with this is continued deforestation and increased regional precipitation caused by changing climate patterns (as already highlighted in Chapter 1). It is, therefore, not only predicted that the frequency of landslides will increase, but the amount of infrastructure and the number of people exposed to their consequences will also increase.

A standard method for classifying landslide types is critical in predicting how they will behave to allow risks to be assessed. The most commonly used classifications of sub-aerial slope movements are based principally on morphology with some account taken of mechanism, material and rate of movement. The type of material is one of the most important factors in the descriptions, due to its influence on the behaviour of a landslide. Figure 7.3 illustrates the various types of movement-material combinations for the fundamental landslide classifications originally defined by Varnes (1978). The following provide brief definitions of the movement types:

- *Fall* – material detaches from a steep slope with little or no shear displacement and descends through the air.
- *Topple* – the resultant vector of applied forces falls through or outside a pivot point causing material to rotate forwards.
- *Rotational slide* – sliding on one or more concave-upward failure surfaces.
- *Translational (planar) slide* – sliding on a planar failure surface running approximately parallel with the slope (i.e., ground surface).
- *Spread* – material fracturing and lateral extension due to liquefaction or plastic flow of subjacent material.
- *Flow* – disaggregation of material with motion comparable to that of a fluid over a rigid bed where the velocity distribution varies with depth (i.e., minimum velocity at the base due to frictional resistance from the basal surface).
- *Complex slide* – slides involving a combination of two or more of the fundamental movement types.

Different types of landslides experience different behaviour and this should be taken account of in risk management. Some landslides can fail slowly over a period of hours, days or weeks, and not reach high rates of movement or large displacements. However, other types of landslides can fail suddenly, experience high rates of movement, and travel large distances. Consequently, the risk reduction and management strategies used for each slope must be different (see Case Study 7.1).

7.4   Causes and Impacts of Landslides

Slope stability is principally dictated by the ratio of shear strength (i.e., stabilising forces) and shear stress (i.e., destabilising forces) respectively along the shear surface(s) (i.e., at the interface between the failing mass and the stable earth beneath). Therefore, landslides occur as a result of a reduction in shear strength (within the soil and/or rock) or an increase in applied shear stress. Shear strength reduces in two main ways: through structural changes as a result of processes such as weathering; or an increase in pore pressures (i.e., the pressure in the pore spaces within the soil and/or rock). Pore pressures increase after periods of rainfall, and in certain materials they increase significantly during shaking (i.e., during an earthquake), which causes liquefaction. Shear stress can increase in many

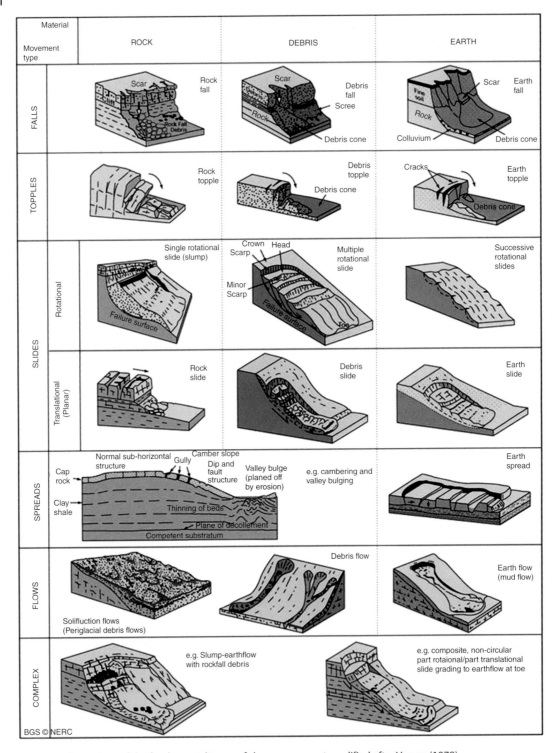

Figure 7.3 Illustrations of the fundamental types of slope movement modified after Varnes (1978) (*Source:* Reproduced with permission of BGS).

CASE STUDY 7.1

The Gansu Landslide

On August 7, 2010, in the north-west Chinese province of Gansu, rainfall-induced debris flows devastated the small county of Zhouqu. The debris flows killed more than 1,700 people and destroyed more than 5,500 properties. A number of factors are reported to have contributed to the occurrence of these debris flows: intensive rainfall following a period of drought, which infiltrated the surface cracks that formed during the dry period; the 7.9 Mw magnitude Wenchuan earthquake of 12 May 2008, which reduced the strength of the material on the hillside; and long-term deforestation.

Prior to this catastrophic event, debris flow hazard had been recognized in the region, but its potential for such devastating impacts was not fully appreciated. This event has motivated organisations to improve debris flow risk management. It has been recommended to put restrictions on land use in such hazardous areas and to relocate at-risk people to safer places. Hazard maps are also being produced to be used by planners of the Department of Land and Resources of Gansu Province and other agencies to inform new land use policies. In addition, recommendations on the design of early warning systems and emergency protocols are ongoing.

Useful search terms:
Gansu, Zhouqu, Wenchuan, landslide, debris flow

Thinking Point 7.2

Avalanches

Although avalanches are not technically landslides, they can be classified with landslides under the term 'mass movements'. In mountainous terrain, avalanches present a significant hazard to life and property because large volumes of snow can travel great distances at high speeds. They can also block roads and railways, and cut off power lines (Figure 7.4).

Avalanches (or snowslides) are typically triggered by a mechanical failure in the snow pack when the stress exceeds the snow pack's strength. After they initiate, avalanches typically accelerate rapidly and grow in volume as they entrain more snow. They can fail due to weakening in the snowpack or increased load due to precipitation, or by seismic activity. They can also be triggered by human (e.g., skiing) or animal activity. In fact, a significant percentage of people killed by avalanches trigger them themselves, and then they typically die from lack of oxygen when buried in the snow.

Avalanche control activities are conducted to reduce the risk they pose. Risk is initially assessed using terrain surveys and surveys of adjacent infrastructure and property. Avalanche control techniques either directly intervene in the evolution of the snow pack, or lessen the effect of an avalanche once it has occurred. This can be achieved by triggering less hazardous avalanches and disrupting weak layers using mechanical or explosive methods. Other techniques include snow retention structures (e.g., racks, nets and barriers). Avalanche control organisations have response and recovery plans in the event of an avalanche, and monitoring and warning systems are being used more widely to detect avalanche evolution.

Useful search terms:
Avalanche, snowslide, snowslip, snow, avalanche control

Figure 7.4 Avalanche (*Source:* The Cabin On The Road on Flickr).

ways, such as: removal of lateral or underlying supports (e.g., due to excavation or erosion); surcharge (e.g., constructing on the top of the slope); earthquakes and other forms of transitory earth stress; altered lateral earth pressures; and volcanic process. The most common causes of landslides globally are prolonged and/or intensive rainfall, and earthquakes.

Many natural slopes fail as a consequence of their stability reducing with time (due to natural phenomena) or in response to extreme rainfall and/or earthquakes. However, many natural slopes fail as a result of human activity, and engineered infrastructure slopes fail as a result of inadequate design and/or construction and/or maintenance. Moreover, some communities continue to build their homes and infrastructure below and across marginally stable slopes.

The cost of emergency repair in response to slope failures can be 10 times greater than the cost of planned maintenance if conducted prior to collapse, and this highlights the importance effective monitoring to detect slope instability prior to failure; however, inadequate monitoring systems are often implemented, if any are used at all.

Examples of human activities that contribute not only to landslide occurrence, but also to landslides interacting with people and infrastructure when they occur, include:

- Rapid urban expansion and inadequate urban planning, which leads to homes and infrastructure being constructed on or beneath slopes with landslide potential, and can lead to construction activity reducing the stability of slopes;
- Inadequate design and/or construction of engineered slopes to resist, for example, climate effects (and forecasted climate change). An example includes the UK railway earthworks that were built in the early- to mid-nineteenth century before modern soil mechanics theories, with little or no knowledge of the processes that can occur after construction;
- Deforestation of slopes, which changes their hydrology, potentially reducing their strength and increasing their vulnerability to rainfall and/or earthquakes;

- Inadequate slope monitoring and early warning strategies, which should be used to understand transient changes to slope stability, enable timely repair and maintenance, and evacuate people; and
- Lack of, or ineffective, emergency preparedness procedures.

Examples of primary effects of landslides include:

- Loss of life;
- Damage to built environment infrastructure;
- Loss of property and homelessness;
- Breakdown of basic services such as water supply and transport;
- Train derailments;
- Collapse of water retaining structures, which can lead to flooding; and
- Rupture of buried pipelines.

Examples of secondary effects of landslides include:

- Disruption to the local and national economy;
- Psychological damage in response to loss of life, serious injuries and/or loss of property;
- Public outrage and social upheaval;
- Effect on reputation and business confidences; and
- Potential litigation

CASE STUDY 7.2

The Hatfield Colliery Landslide

A landslide occurred on a spoil heap used by the Hatfield Colliery, UK, in February 2013, following a period of above average precipitation. The landslide caused significant damage to adjacent rail infrastructure, which had to be closed for several months while it was remediated (Figure 7.5). The landslide distorted large lengths of track on the Doncaster to Goole and Doncaster to Scunthorpe lines, and also affected a major freight line.

Remediation comprised the movement of approximately 1 million cubic metres of material, and replacement of approximately 500 metres of railway. The line reopened in July 2013, however, aspects of the remediation effort continued beyond this. It is fortunate this event did not result in loss of life; however, the event resulted in significant economic loss and impacted severely on the UK's transport infrastructure.

Useful search terms:
Hatfield colliery landslide, Network Rail

7.5 Risk Management

7.5.1 Hazard Identification

Landslide hazard identification can be performed in numerous ways. These can be categorised into three principal groups (Table 7.3): desk study; site inspection; and detailed investigation. It is typically the engineer (geotechnical engineer or engineering geologist) who will carry out the hazard identification process.

Figure 7.5 The Hatfield colliery landslide (*Source:* Reproduced with permission of BGS).

Table 7.3 Sources of Information for Landslide Hazard Identification.

	Sources of information for hazard identification
Desk study	• Existing susceptibility and hazard (i.e., landslide potential) maps and vulnerability and risk zonation maps, particularly for natural slopes, which can often be obtained from government bodies, local authorities or research bodies (e.g., the British Geological Survey in the UK). • Earthworks asset owners (e.g., transport infrastructure) often keep records of their slopes, which include their age, method of construction, performance history, the performance of adjacent and similar earthworks, and often have their own in-house risk maps for their engineered slopes. • Published geological maps (and now online interactive geological maps) provide information on the material type, topography, landslide history and any fault locations for an area of interest. • Topographical information can be obtained from contour maps or computer models developed from previous surveys (e.g., digital elevation models).
Visual inspection/ field observation	• Collect information necessary to perform slope stability analyses (e.g., slope geometry). Such observations can be used to predict the volume/mass, travel distance and velocity of potential landslides. • Identify potential landslide types and mechanisms (e.g., debris flow, rock fall or translational slide) to assess the hazard and its significance. • Identify features indicative of historical and/or recent slope movement, for example: tension cracks at the ground surface; patch repairs to a cracked road surface; trees on a slope with a concave upward shape. • Identify if any other landslides have occurred in the area, and if any slope stabilisation measures have been put in place.

(Continued)

Table 7.3 (Continued)

Sources of information for hazard identification
Detailed investigation	• Slope stability analyses to assess likelihood and type of failure. • Ground investigations provide detailed information on ground water and material properties, which can improve the accuracy of the slope stability analyses. • Monitoring instrumentation can be used to provide spatial and temporal information of the slopes behaviour.

The **desk study** can access: historical information (e.g., existence of ancient landslides or earthwork performance history); existing susceptibility and hazard (i.e., landslide potential) maps; geological information; and topographical information. This allows existing landslides and slopes with landslide potential to be identified. The **site inspection** can reveal features associated with old and new landslide movements (e.g., tension cracks and fallen boulders) and impact on any adjacent infrastructure (e.g., road surface cracks in Figure 7.6). The **detailed investigation** includes slope stability analysis, which accurately models the stability of the slope to assess if it is likely to fail. The accuracy

Figure 7.6 Cracking on road surface due to slope movement (*Source:* Oregon Department of Transportation).

of the slope stability analysis can be increased through ground investigations to obtain material properties and the ground water regime. Monitoring instrumentation can also provide valuable information for input into stability analysis, such as the depth to any existing shear surface, and provides useful information on the slopes behaviour. All this information can be used to produce new landslide hazard and risk maps for the areas of interest, and these can be produced using Geographical Information Systems (GIS). The information collected during the hazard identification process enables the engineer to assign a level of hazard associated with landslide potential. The results from the hazard identification process should inform the decisions made in planning developments and transport routes, and they should impose limitations on land use to restrict development in areas with unacceptable landslide potential. An example map of landslide potential is shown in Figure 7.7, which was produced by the British Geological Survey for the UK.

7.5.2 Assessment of the Vulnerabilities

Vulnerabilities include elements exposed to potential landslide hazard. Such vulnerabilities typically include people, property, transport infrastructure and the built environment, communications infrastructure, utilities and services. High levels of vulnerability are associated with people and infrastructure located on or directly beneath a marginally stable (i.e., could fail any time) slope with a large predicted landslide volume/mass, travel distance and velocity. Table 7.4 provides examples of landslide vulnerabilities, which are categorised using the classifications introduced in Chapter 2.

7.5.3 Determination of the Risk

Many approaches exist for the quantification of risk associated with landslides, and different approaches are used in different countries and organisations (several examples can be found in the books and articles listed at the end of this chapter). However, the fundamental principles are generally applied throughout. Initially, hazards are identified (as explained in the 'Hazard Identification' section) using approaches such as hazard maps and slope stability analyses. The consequences and vulnerabilities of these hazards are then analysed, alongside the frequency of potential landslide events. Risk is then quantified as a function of the likelihood of a landslide, the probability of spatial impact (e.g., travel distance and location of property and infrastructure), temporal probability, vulnerability and the elements at risk (e.g., number of people or value of property and infrastructure). The risk levels obtained are then compared to levels of tolerable or acceptable risk in order to assess priorities and options, and it is usually up to the client, owner and/or regulator to decide to accept the risk, with advice from technical specialists. Risk reduction options are then investigated.

7.5.4 Identification and Prioritisation of Risk Reduction Options

A number of landslide risk reduction options are available, and the selection will depend on the level of the risk in relation to the acceptable/tolerable level, which will be investigated alongside cost-benefit analyses. Chapter 2 introduced the categories of risk reduction; this section will provide specific landslide risk reduction options (Table 7.5).

In general, the risk management options include:

- Acceptance (i.e., the risk is considered within an acceptable range and the project can go ahead);
- Avoidance (i.e., the risk is too high and the project should be abandoned, or an alternative site or revised design should be considered to reduce the risk to within the tolerable range);
- Reduce the likelihood (e.g., increase the resistance of the slope to landsliding);

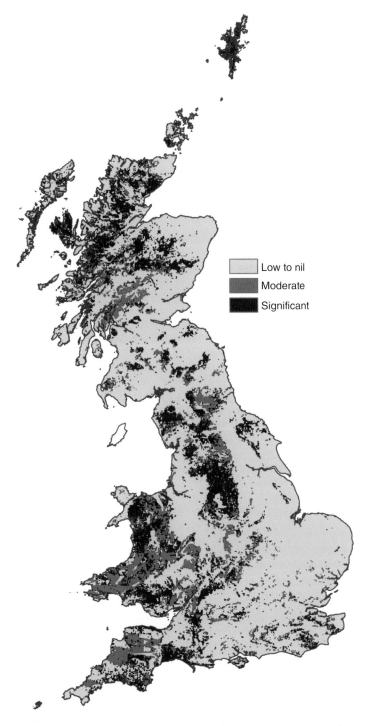

Figure 7.7 Landslide susceptibility map of Great Britain derived from the British Geological Survey GeoSure model (*Source:* Reproduced with permission of BGS).

Low to nil

Moderate

Significant

Table 7.4 Vulnerabilities Associated with Landslides.

Type of vulnerability	Example related to landslide hazards
Physical vulnerability	• Located in areas with high topographical relief and high population density. • Construction materials and machinery are inadequate to achieve a sustainable and resilient built environment. • Rivers below slopes that could be dammed if a landslide occurs, which could eventually break and lead to flooding downstream. • Utilities (e.g., buried pipes) passing through a potential landslide could lead to contamination if damaged. • Location of building/infrastructure in relation to the potential landslide. • Construction activity and utilities/services failure/malfunction (e.g., leaking water pipe) can trigger slope instability. • Remoteness of settlement.
Social vulnerability	• Inadequate awareness to understand slopes and the potential for landslides. • Lack of guidance and regulations on how to take into account landslide risk in the design and construction of built environment infrastructure. • Inability for some individuals (e.g., young children and the elderly) to evacuate in a timely manner in the event of a landslide. • Little choice over where to construct a home (i.e., if beneath/on a potential landslide site is the only option). • Inadequate risk reduction strategies.
Economic vulnerability	• Unable to afford adequate design, construction and remediation, and monitoring and early warning systems. • Unable to afford response mechanisms to evacuate, rescue and remediate. • Lack of economic status leading to little choice over where to construct a home.
Environmental vulnerability	• Located in an area susceptible to prolonged and/or intensive rainfall, and earthquakes. • Utilities (e.g., buried pipes) passing through a potential landslide could lead to contamination if damaged. • Deforested slopes have an increased landslide potential.
Governance vulnerability	• Lack of guidance and regulations on how to take into account landslide risk in the design and construction of built environment infrastructure. • Lack of understanding of how to respond to landslide events. • Inadequate infrastructure and services in place to respond to landslide events. • Inadequate risk reduction strategies.

Table 7.5 Summary of the Viability of Risk Reduction Options for Addressing Landslide Risk.

	Type of risk reduction option for landslide risk					
	Inherent safety	Prevention	Detection	Control	Mitigation	Emergency response
Small volume, low velocity and short travel distance	Y	Y	Y	Y	Y	Y
Large volume, high velocity and large travel distance	#	#	Y	#	#	Y

Please note:
'Y' – indicates that there are possibly a range of useful risk options available
'#' – indicates that some risk reduction options can be used but they are likely to be only of limited effectiveness
'N' – indicates that other than relocating the built asset there is little that can be done to reduce this hazard/threat

- Reduce the consequences (e.g., defensive measures, relocation of the development and/or monitoring and warning systems);
- Transfer the risk (e.g., to another authority); and
- Postpone the decision (e.g., where the uncertainty of the risk is high and further information is required to increase confidence in the decision).

The most cost-effective solution or combination of solutions to achieve satisfactory levels of risk should be selected. Note that certain risk reduction options fall into multiple categories (e.g., stabilisation measures) because their effectiveness depends on many factors, such as the size of the landslide. Table 7.6 provides examples of risk reduction options.

Table 7.6 Indicative Examples of Risk Reduction Options for Addressing Landslide Risk.

Risk reduction option	Examples
1) Inherent safety – eliminate the possibility of hazards occurring	• Do not build on or beneath slopes with significant landslide risk.
2) Prevention – reduce the likelihood of hazards	• Stabilisation measures to control the initiating circumstances, such as re-profiling the surface geometry (e.g., to reduce the angle of the slope), optimise vegetation (e.g., to increase the slope's strength), groundwater drainage (e.g., to increase the slope's strength), anchors and stabilising structures (e.g., pile walls).
3) Detection – measures for early warning of hazards	• Develop rainfall thresholds above which landslides are likely to occur in the slope/area, and compare current and forecasted rainfall information with the thresholds to predict when a slope is likely to fail. • Install deformation and ground water monitoring instrumentation. If these monitor continuously and are connected to a telemetry system, early warning of elevated ground water levels and accelerating slope movements can be used to trigger timely repair and maintenance and/or evacuation of vulnerable people. • In remote and undeveloped areas, a community-based observation method that employs a group of villagers to visually monitor slopes, after being taught basic landslide knowledge, can be used to warn of any signs of instability.
4) Control – limiting the size of the hazards	• Stabilisation measures to control the initiating circumstances, such as re-profiling the surface geometry (e.g., to reduce the angle of the slope), optimise vegetation (e.g., to increase the slope's strength), groundwater drainage (e.g., to increase the slope's strength), anchors and stabilising structures (e.g., pile walls).
5) Mitigation and adaptation – protection from the effects of hazards	• Protective structures (e.g., catch fences and netting). • Relocate development to reduce risk. • Construct accumulation zones for debris, and debris ditches alongside road and rail infrastructure.
6) Emergency response – planning for evacuation emergency access	• Land use zoning. • Emergency planning and 'Local resilience forums'. • Evacuation route planning. • Locating critical infrastructure and emergency services/resources in safe/protected areas. • In remote and undeveloped areas, a community-based observation method that employs a group of villagers to visually monitor slopes, after being taught basic landslide knowledge, can be used to warn of any signs of instability.

Figure 7.8 A catch fence installed along a slope to prevent debris and rocks falling onto the road (*Source:* emdot on Flickr).

An example of a catch fence being used to prevent rocks and debris falling onto a road is shown in Figure 7.8, and an example of an anchored pile wall installed to stabilise a slope that was causing damage to a road is shown in Figure 7.9.

Numerous conventional and emerging slope monitoring technologies offer the potential for early warning, which would enable evacuation of vulnerable people and timely repair and maintenance. Such effective early warning systems monitor deformations continuously, as this informs on whether

Figure 7.9 An anchored pile wall installed to stabilise a slope that was causing damage to a road in Peace River, Canada (*Source:* Alister Smith).

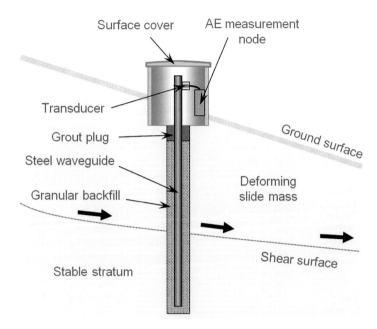

Figure 7.10 The Slope ALARMS landslide early warning system (after Smith *et al.*, 2014 and Dixon *et al.*, 2015).

the slope is moving, how fast it is moving, and whether it is accelerating. All of this information is extremely useful in providing an early warning.

An emerging technology that provides a means of early warning of slope instability is the Slope ALARMS system (Figure 7.10). The system is installed in a hole through the slope, and as the slope moves, acoustic signals are generated by deformation of the gravel around the waveguide. These acoustic signals are related to how fast the slope is moving. The Slope ALARMS monitoring system then sends a text (SMS) message to responsible persons to warn them if the slope is accelerating.

7.6 Summary

Landslides are caused by a complex mix of natural and human-induced processes. Examples of human activities that contribute not only to landslide occurrence, but also to landslides interacting with people and infrastructure when they occur, include:

- Rapid urban expansion and inadequate urban planning;
- Inadequate design and/or construction of engineered slopes to resist, for example, climate effects (and forecasted climate change);
- Deforestation of slopes, which changes their hydrology, potentially reducing their strength and increasing their vulnerability to rainfall and/or earthquakes;
- Inadequate slope monitoring and early warning strategies, which should be used to understand transient changes to slope stability, enable timely repair and maintenance, and evacuate people; and
- Lack of, or ineffective, emergency preparedness procedures.

CASE STUDY 7.3

The Abe Barek Landslide

On the morning of May 2, 2014, in Ago District, Badakhshan Province, Afghanistan, a landslide buried 86 houses and killed almost 2,700 people. The landslide is reported to have occurred in two stages. The first buried numerous houses in Abe Barek village, and the second event occurred two hours later burying rescuing villagers and even more property (Figure 7.11). Farming and irrigation activity, and historical landslide activity, may have contributed to the occurrence of this event, however, prolonged rainfall leading up to the event is anticipated to have been the trigger.

Other landslide sites can be identified in the mountainous area and the steep slope on which this landslide occurred had features indicative of landslide potential. However, little research and investigation was conducted prior to the event because of the unstable political and economic conditions in the area, which meant that effective risk management strategies were not implemented.

If adequate risk management was implemented and monitoring and early warning systems in particular had been put in place, vulnerable people may have been able to evacuate prior to this catastrophic event.

Useful search terms:
Abe Barek landslide, Afghanistan landslide

These human influences on whether a landslide hazard becomes a disaster or not suggests that there is quite a lot that construction practitioners, home owners and individuals can do to reduce the likelihood of a landslide from occurring, well at least for smaller scale landslides.

Figure 7.11 The Abe Barek landslide (*Source:* Reproduced with permission of International Organization for Migration).

Thinking Point 7.3

The Oso Landslide – Could the Risks Have Been Reduced?

A steep hillside at Oso, Washington State, USA, had a history of landslides. It was in an area with high annual rainfall and ongoing deforestation, and the toe of the slope was being actively eroded by the Stillaquamish River. These factors highlighted the landslide potential and were documented; scientists drew attention to the hazard and authorities had listened historically, to some extent, restricting logging activities on occasion. Some defensive structures (e.g., embankments and walls) were even built, but they were destroyed by landslides decades ago. Despite this, the town continued to issue building permits and developments grew.

On the 22 of March 2014, a huge landslide killed 44 people at the bottom of the hillside at Oso (Figure 7.12). The size and behaviour of the landslide was not quite what scientists predicted; it was highly mobile and this is why it was so catastrophic. Scientists studying the landslide suggest that the high mobility could be due to a phenomenon called 'base liquefaction'. There is also no clear trigger of the landslide as there were many contributing factors, however, heavy rainfall prior to the event is likely to have played is a significant role.

The hazards associated with the slope were known but inadequate risk reduction methods were implemented, and this resulted in significant loss of life and damage to the town. What risk reduction measures could have been implemented? Significant expense would have been required to attempt stabilisation measures, in the form of drainage systems and pile walls, because the slope was large. Development beneath the hill could have been restricted and existing property relocated, although this would also have high expense. It is possible that monitoring instrumentation installed across the slope, in conjunction with pre-established response strategies, could have allowed early warning to enable evacuations.

Useful search terms:
Oso landslide, Washington State, United States Geological Survey

---

Key Points:

- Thousands of people are killed globally each year by landslides. They damage built environment infrastructure, which costs billions of pounds to repair, they leave thousands of people homeless, and they cause the breakdown of basic services such as water supply and transport.
- The most common causes of landslides globally are prolonged and/or intensive rainfall, and earthquakes.
- There are numerous approaches to reduce the risk from landslides, such as stabilising the slope to reduce the hazard, or building protective structures to restrict the landslide impacting on people and infrastructure. However, where the scale of the landslide is large and property has not been relocated, monitoring and early warning systems may be the only option; to enable evacuation of vulnerable people.

Reinforcement

1) What are the main triggers of landslides?
2) What sources of information can be used for landslide hazard identification?
3) What landslide risk reduction options are available for 'prevention'?
4) What landslide risk reduction options are available for 'detection'?
5) What landslide risk reduction options are available for 'mitigation and adaptation'?

Figure 7.12 The Oso landslide (*Source:* U.S. Geological Survey).

Questions for discussion

1) How can we enforce the importance of landslide risk management in remote and undeveloped areas?
2) What risk reduction options could have been used to reduce the effects of the landslide case studies and talking points?
3) Why were inadequate risk reduction options implemented in the case studies and talking points? What are the barriers for their implementation in projects in general? Who is required to pay for the risk reduction options?

Further Reading

Books:

Anderson MG & Holcombe E. (2013). *Community-based Landslide Risk Reduction: Managing Disasters in Small Steps*, World Bank Publications

Cornforth, D. (2005). *Landslides in practice: investigation, analysis, and remedial/preventative options in soils*, Wiley.

Dunnicliff J. (1993). *Geotechnical instrumentation for monitoring field performance*, John Wiley & Sons.

Glade T., Anderson M.G. and Crozier M.J. (Eds.), (2006). *Landslide hazard and risk*, John Wiley & Sons

Highland L. and Bobrowsky P.T. (2008). *The landslide handbook: a guide to understanding landslides*, Reston, VA, USA: US Geological Survey

Sidle R.C. and Ochiai H. (2006). *Landslides: processes, prediction, and land use*, American Geophysical Union

Articles:

Cruden D.M. and Varnes D.J., (1996). 'Landslides: Investigation and Mitigation. Chapter 3-Landslide types and processe', *Transportation research board special report*, (247)

Dai F.C., Lee C.F. and Ngai Y.Y. (2002). 'Landslide risk assessment and management: an overview'. *Engineering Geology*, 64(1), pp. 65–87

Dixon N, Smith A, Spriggs MP, Ridley A, Meldrum P & Haslam E (2015). 'Stability monitoring of a rail slope using acoustic emission'. *Proceedings of the Institution of Civil Engineers: Geotechnical Engineering*, 168(5), pp. 373–384

Gibson AD, Culshaw MG, Dashwood C & Pennington CVL (2013). 'Landslide Management in the UK - the Problem of Managing Hazards in a 'Low-Risk' Environment', *Landslides*, 10(5), 599–610

Glendinning S, Hall J and Manning L (2009). 'Asset-management strategies for infrastructure embankments', *Proceedings of the Institution of Civil Engineers: Engineering Sustainability*, 162(2), 111–120

Guzzetti F, Carrara A., Cardinali M. and Reichenbach P., (1999). 'Landslide hazard evaluation: a review of current techniques and their application in a multi-scale study, Central Italy', *Geomorphology*, 31(1–4), 181–216

Hungr O., Leroueil S. and Picarelli L. (2014). 'The Varnes classification of landslide types, an update', *Landslides*, 11(2), 167–194

Iverson R.M., George D.L., Allstadt K., et al. (2015). 'Landslide mobility and hazards: implications of the 2014 Oso disaster', *Earth and Planetary Science Letters*, 412, 197–208

Perry J., Pedley M. and Reid M. (2003a). *Infrastructure embankments – condition appraisal and remedial treatment, C592*. London: Construction Industry Research and Information Association (CIRIA), London

Perry J., Pedley M. and Reid M. (2003b). *Infrastructure cuttings – condition appraisal and remedial treatment, C592*. London: Construction Industry Research and Information Association (CIRIA), London

Petley D.N. (2012). 'Global Patterns of Loss of Life from Landslides'. *Geology, Geological Society of America* 40 (10): 927–930

Smith A., Dixon N., Meldrum P., Haslam E. and Chambers J. (2014). 'Acoustic emission monitoring of a soil slope: Comparisons with continuous deformation measurements', *Géotechnique Letters* 4(4), 255–261

Tang C., Rengers N., van Asch T.W., Yang Y.H. and Wang G.F. (2011). 'Triggering conditions and depositional characteristics of a disastrous debris flow event in Zhouqu city, Gansu Province, northwestern China'. *Natural Hazards and Earth System Science*, 11(11), 2903–2912

Uhlemann S., Smith A., Chambers J.E., Dixon N., Dijkstra T., Haslam E., Meldrum P.I., Merritt A.J., Gunn D.A. and Mackay J. (2016). 'Assessment of ground-based monitoring techniques applied to landslide investigations'. *Geomorphology*, 253: 438–451

Van Westen C.J., Van Asch T.W. and Soeters R. (2006). 'Landslide hazard and risk zonation—why is it still so difficult?', *Bulletin of Engineering geology and the Environment*, 65(2), 167–184

Varnes D.J. (1978). 'Slope movement types and processes', *Transportation Research Board Special Report, (176)*

Useful websites:

www.slopealarms.com

http://www.bgs.ac.uk

http://www.usgs.gov/

http://blogs.agu.org/landslideblog/

Section IV

Key Considerations and Ways Forward

8

Key Principles

It is urgent and critical to anticipate, plan for and reduce disaster risk in order to more effectively protect persons, communities and countries, their livelihoods, health, cultural heritage, socio-economic assets and ecosystems, and thus strengthen their resilience.
Sendai Framework for Disaster Risk Reduction, 2015–2030

It has already been demonstrated in the previous chapters that the impacts of natural hazards and human-induced threats on the built environment can wipe out years of long-term investment in development of infrastructure and associated buildings. Economic losses from disasters such as earthquakes, tsunamis, cyclones and flooding are now reaching an average of US$250 billion to US$300 billion each year. Future losses (expected annual losses) are now estimated at US$314 billion in the built environment alone, with the growth in losses being on the increase as a result of demographic, economic and socio-political challenges including urbanisation, climate change and terrorism threat.

Rapid development of the built environment in recent decades has too often taken place with little regard for the underlying natural hazards let alone changing climatic conditions and the dynamic socio-economic and political pressures associated with mass urbanisation and poorly regulated development. As a consequence, many of the disasters being encountered in recent decades have tended to be a predictable result of the interactions between the natural and built environments and the communities that reside and work within these environments.

8.1 Learning Objectives

By the end of this chapter you will learn:

- Why it is important to implement both structural and non-structural approaches to DRR
- What role the construction sector and other stakeholders can play in DRR
- What principles should be considered when implementing DRR in the built environment

8.2 Integrating DRR Measures into Construction Practice

As discussed throughout this book, there is an increasing threat to the built environment from a variety of different hazards and threats, some of which are well-known but hard to mitigate, and others that can be emergent and thus less predictable. Due to changes in the climate, a large number of hazards and threats are likely to become more frequent and significant in the near future; it is thus

Disaster Risk Reduction for the Built Environment, First Edition. Lee Bosher and Ksenia Chmutina.
© 2017 John Wiley & Sons Ltd. Published 2017 by John Wiley & Sons Ltd.

Figure 8.1 The Thames Barrier, London (*Source:* Damien Moore on Flickr).

critical for those involved in planning, design, construction and operation of the built environment to take serious consideration of these hazards and threats as a core part of their professional activities. The decisions taken now will determine the burden that future generations inherit with regards to their resilience to a range of hazards; so efficient planning, design and construction today should lessen the need for expensive retrofitting measures in the future.

8.2.1 Resilient Built Environment

A **resilient built environment** should be designed, located, built, operated and maintained in a way that maximises the ability of built assets, associated support systems (physical and institutional) and the people that reside or work within the built assets, to withstand, recover from, and mitigate for, the impacts of natural hazards and human-induced threats.

The idea of built-in resilience is a key component of DRR. To design a resilient built environment means to start thinking carefully during the early design process about the typical use scenarios of the city; particularly considering the common points of stress due to normal use, as well as the most likely disaster situations that could challenge the integrity of the city and/or endanger its inhabitants and functions. The local environment always plays a critical role in determining the factors that make a city resilient or not, and so resilient design is always locally specific (see Thinking Point 8.1).

It is important to ensure the system or built asset is an improved (i.e., more robust or resilient) version of what was originally there (this is particularly the case for assets that have been historically at risk from hazards such as flooding, windstorms, earthquakes and landslides). If built assets are repeatedly affected by particular hazards, we need to learn lessons from this and replace the original structure(s) with an improved version that is more resilient (in social, physical and economic terms).

Thinking Point 8.1

Resilience of What To What?

Resilience is hard to define; with the definitions often changing depending on the context in which resilience is being viewed. Approaches to resilience tend to provide different importance to the objectives of avoidance (avoid the shock), recovery (rebound after the shock) and withstanding (resist the shock). There are three common models in the adoption of resilience:

- the mitigation model, which emphasises a reduction of exposures and risks;
- the recovery model, which accepts that not all the shocks can be eliminated and thus embraces actions that are required after them;
- and the structural-cognitive model, which focuses on structural changes in society and institutions, the importance of situational factors (physical location, age, income, etc.) and cognitive factors (psychological and attitudinal).

Another debate is about the level at which resilience must be achieved. Much of the discussions can be grouped into the following categories:

- resilience of the relationships between societies and the natural environment;
- local resilience, including urban resilience, local and community level adaptation to climate;
- holistic or a broader systems approach to resilience as a basis for disaster risk reduction.

Taking all this into account, it can be seen that resilience integrates:

- multiple levels of analysis and intervention (from individual to the national level);
- multiple time-scales: prevention, emergency, rehabilitation, reconstruction, long-term development;
- multiple sectors of intervention, including emergency action, construction industry development and housing development; and
- multiple types of intervention and units of analysis.

The main challenges in achieving built-in resilience lie in the culture and structure of the construction sector, which comprises a complex and temporary array of inter-organisational relationships, governed by project-defined interactions. These challenges include:

- Most construction projects are normally planned, designed and constructed by a combination of companies and individuals, many of whom have not worked together previously and will most likely not work together again.
- Each project is different thus it is impossible to introduce the degree of routinisation and repetition that can be achieved in other industries.
- Construction projects often involve balancing the objectives of the firm with those of the project within a constrained budget.
- The relatively low skill levels required for many construction operations and the low barriers to entry can lead to an informal labour market, which is extremely difficult to regulate.
- Corruption in some parts of the world exacerbates the vulnerability of buildings and structures.
- There is a high level of professional fragmentation, with architects, surveyors and specialist engineers often employed from outside construction firms as independent consultants.

These challenges are deeply ingrained into the construction industry, thus making it extremely difficult to enact structural and cultural changes. Thus, adding expertise in mitigating the effects of disasters to an already cluttered project team complicates the structure of responsibilities.

Thinking Point 8.2

Should Aesthetics Be Compromised?

To design a resilient building means to start the design process by thinking carefully about the typical use scenarios of the building, common points of stress due to normal use, as well as the most likely disaster situations that could challenge the integrity of the building and/or endanger its occupants. It is important to ensure the system or built asset is an improved (i.e., more robust or resilient) version of what was originally there (this is particularly the case for assets that have been historically at risk from hazards such as flooding, windstorms, earthquakes and landslides). But at the same time it is important to think about the fact that resilient features of the built environment should be unobtrusive. This draws on methods popularised in North America in the post 9/11 era of utilising softer, more subtle and 'landscaped' security—what has also been referred to as invisible security—so that security features are designed into ornamental fences, sculptures, large flower planters, or where trees are used as a defensive cordon instead of highly visible security barriers (Jersey barriers). For example, security planning in Washington D.C. in 'The National Capital Urban Design and Security Plan' proposes six key goals, which are an active attempt to avoid 'fortress'-style security and to coordinate better the future development of security policy:

1) Provide an appropriate balance between the need to accommodate perimeter security for sensitive buildings and their occupants and the need to maintain the vitality of the public realm.
2) Provide security in the context of streetscape enhancement and public realm beautification, rather than as a separate or redundant system of components, the only purpose of which is security.
3) Expand the palette of elements that can gracefully provide perimeter security in a manner that does not clutter the public realm, while avoiding the monotony of endless lines of jersey barriers or bollards, which only evoke defensiveness.
4) Produce a coherent strategy for deploying specific families of streetscape and security elements in which priority is given to achieving aesthetic continuity along streets, rather than solutions selected solely by the needs of a particular building under the jurisdiction of one public agency.
5) Provide perimeter security in a manner that does not impede the city's commerce and vitality, excessively restrict operational use of sidewalks or pedestrian and vehicular mobility, or impact the health of existing trees.
6) Identify an implementation strategy that can be efficiently coordinated in the most cost-effective manner.

In the UK the importance of 'unobtrusive' security was exemplified by the Emirates football stadium in North London, which was held up as a model for designing-in counterterrorism to new buildings. Rather than using the typical type of 'bollard' to prevent vehicular access (see Figure 8.2) to a building/site the designers utilised an innovative feature. The stadium is ringed by a variety of ornaments or streetscape designs, from reinforced benches to large brass cannons (Arsenal football club's insignia) and large 'toughened' concrete letters spelling out 'Arsenal' (Figure 8.3), which are deliberately situated to prevent vehicle access and, according to reports, are designed to stop a seven-tonne lorry.

It is, however, important to develop a process within which resilience can systematically be built-in. Such frameworks already exist (e.g., building regulations, procurement practices, design process), but there is no structured way to ensure that this guidance and practice are utilised effectively.

Figure 8.2 Bollards used to restrict vehicular access in Jerusalem (*Source:* Lee Bosher).

Figure 8.3 'Arsenal' sign used to restrict vehicular access to the Emirates Stadium, London (*Source:* Lee Bosher).

8.2.2 Structural and Non-Structural Approaches

As emphasised throughout this book, different approaches can be taken in order to reduce the risk of a disaster, some of which are structural and others are non-structural. Common structural measures for disaster risk reduction include dams, flood levies, ocean wave barriers (see Figure 8.4), earthquake-resistant construction, and evacuation shelters.

Figure 8.4 An example of concrete 'tetrapods' used to reinforce the coastline on Ikei Island, Japan (*Source:* Nelo Hotsuma on Flickr).

Common non-structural measures include building codes, land use planning laws and their enforcement, research and assessment, information resources, and public awareness programmes (see Figure 8.5).

It is important to remember that that in civil and structural engineering, the term 'structural' tends to be used in a more restricted sense to refer to the load-bearing structure, with other parts such as wall cladding and interior fittings being termed non-structural. For further clarification, the definitions of the term structural and non-structural are provided below (Figure 8.6).

Figure 8.5 An example of non-structural measures: earthquake drill at school (*Source:* Jim Holmes).

Structural measures include any physical construction to reduce or avoid possible impacts of hazards, or application of engineering techniques to achieve hazard-resistance and resilience in structures or systems.

Non-structural measures are any measure not involving physical construction that uses knowledge, practice or agreement to reduce risks and impacts, in particular through policies and laws, public awareness raising, training and education.

Figure 8.6 Hafencity: Utilising a masterplan that considers the interaction between existing and new buildings, the waterfront and the elevation of buildings as a flood protection concept (*Source:* Reproduced with permission of M. Pinke on Flickr).

Structural measures typically decrease vulnerability of a building thus reducing the expected loss, however, very often structural and non-structural measures are implemented together (see Case Study 8.1).

The most cost-effective forms of DRR investment tend to be non-structural approaches, including land use planning, warning systems, and household/individual-level life style changes. These approaches often need to be supported by structural measures; for example, for earthquakes, bolting down appliances and securing shelves costs several hundred dollars in order to save several thousand dollars in averting potential damages. That presumes that the entire building will stand up in an earthquake because it has been engineered with seismic safety measures.

It is, however, important to understand that structural measures do not fully reduce the risk: instead of controlling the risk of, for instance flooding, they control the hazard parameters (e.g., the volume of water), which can increase in the future and therefore make structural measures ineffective. It is thus important to strike a right balance in implementing both structural and non-structural measures.

CASE STUDY 8.1

HafenCity

In 2000 Hamburg City decided to transform part of the former Harbour area into a new residential, office and retail area. A master plan was made for 'HafenCity' with the aim to build an attractive living environment closely connected to the water. The area lies in front of the main dike-line of Hamburg adjacent to the Elbe River. The City developed a mix of innovative strategies to manage flood risk, instead of building a dike. They introduced the 'Warftenkonzept' - building on elevated plots in front of the main dyke line with heights of over 7.5 meters (25 feet) above 'Normal Null /sea level' (NN), soon to be upgraded to over 8.3 meters (27 feet) NN for the new neighbourhoods, resulting from new predictions for sea level rise. All newly built infrastructure in HafenCity is elevated to allow access of the fire brigade during storm surges. Individual built-in flood resistance for buildings such as flood doors and walls was introduced, as well as the institutionalisation of 'Flutschutzgemeinschaften' among property owners and inhabitants of particular neighbourhoods in HafenCity. They are responsible for flood preparedness, for timely alert during a flood event and for closing the mobile flood protection walls.

Search terms: HafenCity; Hamburg city; flooding

8.3 Seven Key Principles

As has been demonstrated in the previous sections, there is a need to find a way in which resilience can be more systematically built-in. With the concept of resilience being context specific and the wide range of DRR options on offer, it is typically impossible to develop a single model for building resilience, but there are some broad principles, many of which overlap with each other, that can be a useful point of reference (Bosher and Dainty, 2011).

8.3.1 Principle 1: Adopt a Holistic Perspective

Often disaster risk reduction practices are not applied for the following reasons: the failure to apply knowledge in a suitably integrated way, the failure to agree common standards, the need for professionalism, problems with knowledge transfer, and a disproportionate focus on technological over social solutions. In addition, resilience activities have often been too focused on response (i.e., preparing to respond and getting business back to normal) rather than hazard mitigation and risk avoidance or risk reduction approaches. Resilience and response should be seen as mutually compatible concepts, which allow embedding hazard mitigation and preparedness considerations into post-disaster response and reconstruction and seizing the opportunity to learn and implement lessons learnt (see Case Study 8.2). It is thus critical that a more holistic perspective is adopted where local ('bottom-up') socio-economic and cultural factors are considered alongside the top down, strategic and national considerations. Thus a holistic perspective is one (see Figure 8.7) where:

a) Natural hazards and human induced threats are considered as part of the risk assessment process, and where potential DRR measures are designed to address all the relevant hazards and threats identified;
b) Different disciplines (formally trained and informally skilled) should be provided with opportunities to co-produce solutions that transcend disciplinary siloes;
c) All the phases of disaster risk management (hazard identification, mitigation, preparedness, and recovery/rehabilitation) should be viewed as interconnected and not mutually exclusive;

d) Local coping capabilities, available skills and appropriate technologies should be considered when new technological approaches are being used; and,

e) A deep understanding of the underlying socio-economic and political factors that have resulted in some sections of society being more or less vulnerable to the impacts of disasters, should be included as part of the local risk assessment.

8.3.2 Principle 2: Develop and Appropriately Apply Resilient Technologies

Whilst it is important to develop technologies that could help mitigating threats, it is crucial to bear in mind the socio-economic environment in which this technology is to be used. Too often disaster risk reduction projects are top-down and technology-driven: they may utilise outside engineers and other professionals and use technologies which supplant local knowledge and local labour. In doing so, applying inappropriate technologies and processes is likely to disengage local stakeholders with the development process.

There are a range of examples in this textbook that illustrate some high-tech and low-tech structural solutions. For instance Chapter 3, on flooding, provides some examples of large-scale engineering developments that are protecting millions of people from the impacts of riverine and coastal flooding, that is, the Thames Barrier in London and the Oosterscheldekering surge barrier in the Netherlands. Box 5.2 in Chapter 5 gives some examples of how modern technological developments in materials and structural design have provided some innovative earthquake engineering approaches that are being used to improve seismic performance. In contrast, Case Study 5.4, in the same chapter, gives an insight into traditional earthquake resistant structures in the Himalayas. Such traditional structures that have evolved over many centuries are also sensitive to the locally available resources, the environment in which they are constructed and the spatial requirements of those who inhabit

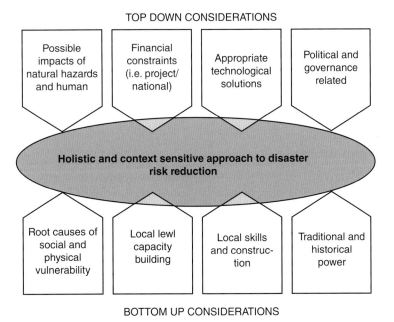

Figure 8.7 Holistic perspective (*Source:* Bosher, 2010).

CASE STUDY 8.2

Rebuilding Manchester After 1996 PIRA Terrorist Attack

The 1996 Manchester bombing was an attack carried out by the Provisional Irish Republican Army (PIRA) on Saturday, 15 June 1996, in Manchester, England. The 1,500-kg bomb, placed in a van on Corporation Street in Manchester city centre, targeted the city's infrastructure and economy and caused widespread damage, estimated by insurers at £700 million (£1.1 billion as of 2015). Although the area was evacuated, the bomb squad was unable to defuse the bomb before it was detonated. Two hundred and twelve people were injured, but there were no fatalities. Several buildings near the explosion were damaged beyond repair and had to be demolished, whereas many more were closed for months for structural repairs (see Figure 8.8).

Such destruction, however, effectively provided a blank canvas for city planners. Much of the 1960s redevelopment of Manchester's city centre was unpopular with residents. Market Street, close to the explosion and at that time the second busiest shopping street in the UK, was considered by some commentators a "fearful" place, to be "avoided like the plague". After the explosion, a number of shops, such as Marks and Spencer, took this opportunity to fully rebuild their damaged stores as well as open new shopping malls. Michael Heseltine, the Deputy Prime Minister at the time, announced an international competition for designs of the redevelopment of the bomb-affected area some of which would include improved security design features. Bids were received from 27 entrants, which led to the creation of the Triangle, Cathedral Gardens, the National Football Museum and other buildings and attractions that gave a boost for the area. Most of the rebuilding work was completed by the end of 1999, at a cost of £1.2 billion, although redevelopment continued until 2005.

Search terms: Manchester: PIRA terrorist attack; Manchester redevelopment

---

them. Arguably, these considerations are pertinent for high income as well as low- or middle-income country contexts

Figure 8.9 illustrates the type of work undertaken by Dhaka Ahsania Mission, a Bangladeshi NGO that was tasked with designing locally appropriate (from an available skills, materials and end-user perspective) toilets and water access points in rural areas that are very prone to flooding from the sea and rivers. Therefore, it is important to reiterate a point made in the previous principle, that local coping capabilities, available skills and appropriate technologies should be considered when new technological approaches are being used.

8.3.3 Principle 3: Engage a Wide Range of Stakeholders (Including Local Communities) in Resilience Efforts

The nature of risks to human development are such that there will inevitably be unintended consequences associated with attempts to improve resilience. It has to be recognised that future risks cannot be predicted any more accurately than the future socio-economic circumstances of the countries in which they will occur, because vulnerability cannot be separated from the social and cultural conditions under which the risks exist. It is thus important to design flexible and responsive solutions which can adapt to changing risks and evolving (and/or devolving) institutional contexts. It is important to advocate for a variety of methods that can help creating various approaches towards meeting local community expectations, supported through meaningful consultation mechanisms. Engaging with the community is a key requirement in the co-production of suitable resilience approaches as well as for the acceptance and long-term ownership of such approaches (see Case Study 8.3).

(a)

(b)

Figure 8.8 Before and after the IRA attack in Manchester (*Source:* Upper: Manchester Fire Service; Lower: Sarah Boulton).

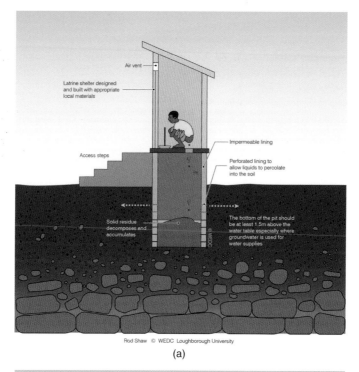

(a)

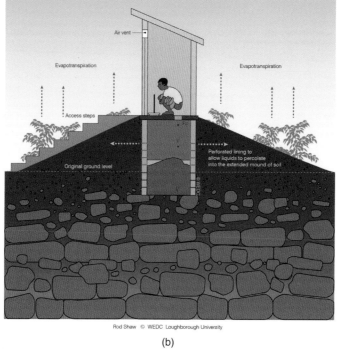

(b)

Figure 8.9 Use of appropriate resilient technologies could include suitable materials and designs for latrines in flood prone areas (*Source:* WEDC after Dhaka Ahsania Mission).

Thinking Point 8.3

Traditional Knowledge for DRR

Even before high technology based early warning systems, or standard operating procedures for response have been introduced, many local communities worldwide have prepared, operated, acted and responded to natural hazards using indigenous methods (many of which are primarily oral) passed on from one generation to the next. Thus it is important to understand, acknowledge and respect indigenous knowledge as a valuable source of information and as a key contributor to reducing risk in many parts of the world.

Indigenous knowledge refers to the methods and practices developed by a group of people from an advanced understanding of the local environment, which has formed over many generations of habitation. This knowledge contains several other important characteristics which distinguish it from other types of knowledge, including originating within the community, maintaining a non-formal means of dissemination, collectively owned, developed over several generations and subject to adaptation, and imbedded in a community's way of life as a means of survival. This knowledge can, for instance, be used to identify early signs of unusual weather patterns or animal behaviour, longer-term impacts on the natural environment as well as insights into emergency evacuation plans and local coping mechanisms.

The relationship between indigenous knowledge and disasters has garnered more interest in recent years. Discussions around indigenous knowledge highlight its potential to improve disaster risk reduction policies through integration into disaster education and early warning systems. Four primary arguments have been made for the value of indigenous knowledge:

- various specific indigenous practices and strategies embedded in the knowledge can be transferred and adapted to other communities in similar situations;
- an incorporation of indigenous knowledge in existing practices and policies encourages the participation of the affected community and empowers its members to take the leading role in all disaster risk reduction activities;
- the information contained in indigenous knowledge can help improve project implementation by providing valuable information about the local context;
- the informal means by which indigenous knowledge is disseminated provides a successful model for other education on disaster risk reduction.

Some useful publications on this topic include:
Shaw, R., Sharma, A. and Takeuchi, Y., 2009. *Indigenous knowledge and disaster risk reduction*. Nova Science Publishers,
Mercer, J., Kelman, I., Taranis, L. and Suchet-Pearson, S., 2010. 'Framework for integrating indigenous and scientific knowledge for disaster risk reduction', *Disasters*, 34(1), pp.214-239

However, on most development projects, the relevant stakeholders will include a very broad range of people from a diverse range of disciplines involved in decision making. Previous research undertaken in the UK has mapped out the main types of stakeholders that should be involved with DRR activities in construction projects and also when their optimal inputs should be made (see Figure 8.10a). The key messages from the matrix shown in Figure 8.10b are:

1) There are many stakeholders that should/could be more involved in DRR related activities; and
2) Specific stakeholder optimal inputs should be targeted at times when the inputs can be most relevant – specifically in the design and planning stages.

CASE STUDY 8.3

Awareness Raising in Barbados

In 2006, the Department of Emergency Management (DEM) was introduced in Barbados. It is responsible for the development and implementation of the Emergency Management Programme and for the coordination of emergency management activities. One of the main responsibilities of the DEM is raising awareness among the general population. It uses mass media, social media, public lectures, exhibitions, open days and videos as well as brochures and publications to reach all segments of the population. The DEM has also been partnering with the public and private sector in using their systems and resources to raise the awareness; for example, the month of June is annually announced as a DRR month when in partnership with various organisations (e.g., Barbados Light and Power (main energy utility) in 2012, and insurance companies in 2013), the DEM organises various events to increase awareness and raise interest in disaster risk reduction before the start of a hurricane season in August/September. The general consensus is that the public's awareness is high, as they are provided with detailed and comprehensive information that can be used to prepare household plans.

Search terms: Barbados, Disaster risk reduction, public awareness, information campaigns

---

This may sound like an obvious statement but many risk managers and emergency planners consulted as part of this research reported that their inputs on any DRR-related matters (i.e., flood defences, sustainable urban drainage systems (SuDS, described in Table 3.7), crime prevention through environmental design tended to be requested far too late in the process (i.e., when everything was planned, designed and the construction was nearing completion).

8.3.4   Principle 4: Utilise Existing Guidance and Frameworks When Appropriate

A key challenge for disaster risk reduction is not in generating new knowledge per se but rather in applying existing knowledge, that is, through the process of technology transfer and diffusion. Existing guidance incorporates elements for building in resilience and can be used as a good starting point for identifying methods of incorporating DRR within professional activities. Many of the existing frameworks are quite flexible thus enabling the users to adjust them for their own requirements and context. For instance, the UN's Sendai Framework for Disaster Reduction has been framed to allow interpretation and adaptation of the contents for local contexts, and the Eurocodes have been developed to encourage the incorporation of national annexes where localised geological and climatic contexts are considered. These are just a couple of examples of the many frameworks, policy and guidance documents that have been developed over recent decades and that can provide useful for adoption and/or adaption to other contexts. The wheel does not have to be reinvented!

The Civil Contingencies Act was developed in the UK in 2004 to establish a coherent framework for emergency planning and response ranging from local to national level activities and explicitly including the inputs of a diverse range of key stakeholders (see Table 8.1). In spite of a number of potential weaknesses in the Act, particularly from a DRR perspective (see Case Study 8.4), it nevertheless provides a good framework for multi-stakeholder engagement and is thus an idea that could be replicated/adapted for other parts of the world.

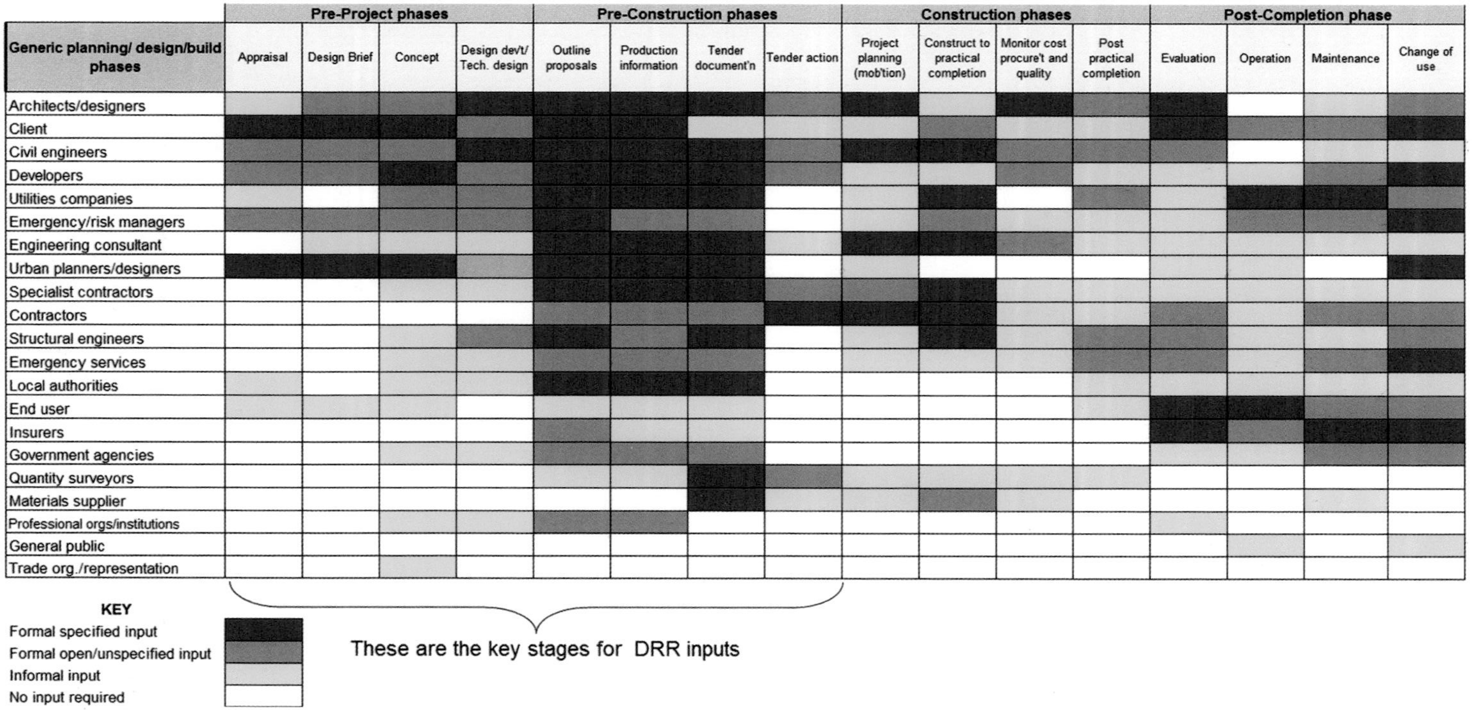

Generic planning/ design/build phases

	Pre-Project phases			Pre-Construction phases					Construction phases					Post-Completion phase		
	Appraisal	Design Brief	Concept	Design dev't/ Tech. design	Outline proposals	Production information	Tender document'n	Tender action	Project planning (mob'tion)	Construct to practical completion	Monitor cost procure't and quality	Post practical completion	Evaluation	Operation	Maintenance	Change of use
Architects/designers																
Client																
Civil engineers																
Developers																
Utilities companies																
Emergency/risk managers																
Engineering consultant																
Urban planners/designers																
Specialist contractors																
Contractors																
Structural engineers																
Emergency services																
Local authorities																
End user																
Insurers																
Government agencies																
Quantity surveyors																
Materials supplier																
Professional orgs/institutions																
General public																
Trade org./representation																

These are the key stages for DRR inputs

KEY
- Formal specified input
- Formal open/unspecified input
- Informal input
- No input required

(a)

Figure 8.10a Stakeholder identification and DRR inputs: Who should be involved and when should the inputs be made (*Source: Bosher, 2014*).

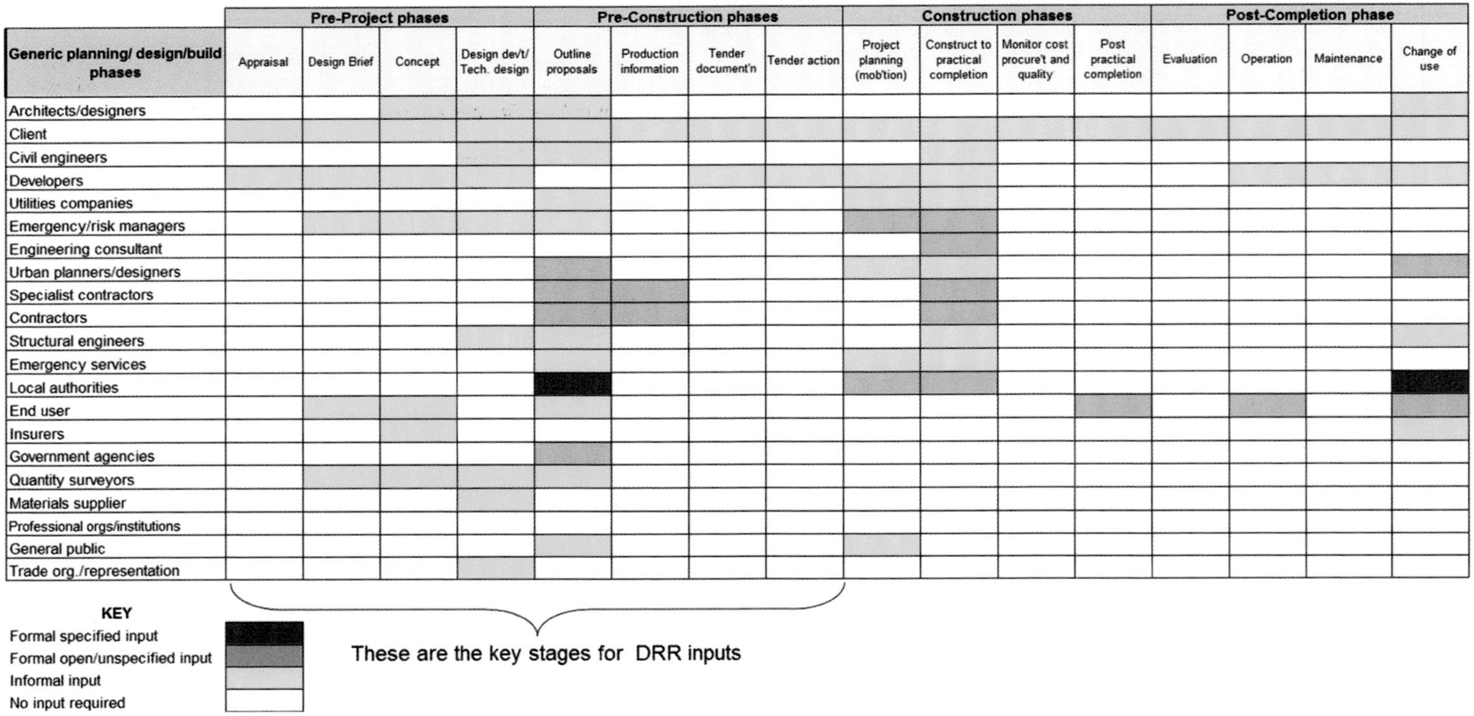

Figure 8.10b Stakeholder identification and DRR inputs

Generic planning/ design/build phases	Pre-Project phases			Pre-Construction phases					Construction phases				Post-Completion phase			
	Appraisal	Design Brief	Concept	Design dev/ Tech. design	Outline proposals	Production information	Tender document'n	Tender action	Project planning (mob'tion)	Construct to practical completion	Monitor cost procure't and quality	Post practical completion	Evaluation	Operation	Maintenance	Change of use
Architects/designers																
Client																
Civil engineers																
Developers																
Utilities companies																
Emergency/risk managers																
Engineering consultant																
Urban planners/designers																
Specialist contractors																
Contractors																
Structural engineers																
Emergency services																
Local authorities																
End user																
Insurers																
Government agencies																
Quantity surveyors																
Materials supplier																
Professional orgs/institutions																
General public																
Trade org./representation																

These are the key stages for DRR inputs

KEY

- Formal specified input
- Formal open/unspecified input
- Informal input
- No input required

(b)

Figure 8.10b Stakeholder identification and DRR inputs: Who is typically involved and when their inputs are usually made (*Source*: Bosher, 2014).

Table 8.1 Organisations ('Responders') Involved with 'Civil Contingencies' in the UK.

CATEGORY ONE ORGANISATIONS
(Category One responders are known as core responders - they include emergency services as well as other important organisations)

Local Authorities	All principal local authorities (County, District, Borough & Metropolitan)
Government agencies	The Environment Agency, the Scottish Environment Protection Agency, Natural Resources Wales,
Emergency Services	Police Forces, British Transport Police, Police Service of Northern Ireland, Fire Services, Ambulance Services, HM Coastguard
National Health Service (NHS) Bodies	NHS hospital trusts, NHS foundation trusts (and Welsh equivalents), NHS England and Public Health England, Port Health Authority

CATEGORY TWO ORGANISATIONS
(Responsible for co-operating with Category One organisations and sharing relevant information)

Utilities	Electricity distributors and transmitters, Gas distributors, Water and sewerage undertakers, Telephone service providers (fixed and mobile)
Transport	Network Rail, Train operating companies (passenger and freight), London Underground, Transport for London, Highways England, Airport operators, Harbour authorities
Others	Health & Safety Executive (HSE), NHS Clinical commissioning group, Voluntary Agencies

8.3.5 Principle 5: Exploit Opportunities to Build-In Resiliency Measures Post-Disaster

It has already been highlighted that there is a need for more resilient measures to be considered post-disaster. In the past, disasters have been dealt with focusing on response activities, that is, saving lives, providing emergency relief and marshalling resources for restoration and reconstruction. In the last few decades it has been recognised that these vital measures should be complimented by a more proactive and developmental approach that addresses pre-event vulnerabilities, and therefore, aims to reduce them in how post-disaster reconstruction is undertaken. It is likely that those responsible for post-disaster reconstruction and retrofitting can be more amenable to taking on board DRR measures, as they have witnessed the effects of the initial disaster (see Case Study 8.5). Also, post-disaster responses and reconstruction efforts that are overly influenced by a political and economic will to 'reconstruct quickly' are not entirely conducive to the longer-term strategies required for the attainment of physical or social resilience.

From a construction perspective, it has also been recognised that addressing the contractual and legislative approaches to post-disaster reconstruction prior to an event can help to build-in resilience for a community by assisting post-event recovery and reconstruction efforts. In addition, the 'Shelter After Disaster Strategies for Transitional Settlement and Reconstruction' (UNOCHA, 2010) was developed to assist all stakeholders responding to rapid-onset disasters, especially those stakeholder responsible for planning and coordination in governments and in humanitarian and developmental organisations. Such frameworks should also be considered for all disaster affected nations as it is these types of proactive measures that are important for building more resilient communities and the critical lifeline services that society so often relies upon.

CASE STUDY 8.4

Civil Contingencies Act in the UK

The UK has an established system for emergency planning and engagement between required stakeholders (see Table 8.1) described in the Civil Contingencies Act (the Act). This system is a network of designated governmental, non-governmental and private sector organisations (typically referred to as 'responders') that can be activated during an emergency and is enacted through exercising and training. This network does not exist permanently (and does not have statutory rights) as the organisations remain formally separate, but is activated if an emergency event occurs. This approach ensures that responders potentially exist at any point in time based on the multi-agency plans that can be changed based on past experiences.

The Act was the starting point of a new contingencies system that has been developed as a result of various events in the period between 1989 and 2001 (including flooding, terrorist incidents, epidemics and infrastructure protests). The overarching aim of the Act is preparedness, so all the decisions in the Act are geared towards the negotiation of the potentialities of the event but at the same opening out the possibilities of response that can be adapted to a specific event.

Whilst the Cabinet Office has ultimate responsibility for civil protection, resilience is carried out through the Local Resilience Forum (LRF) (see Figure 8.11), because emergencies typically start at the local level, and most incidents are expected to be able to be dealt with by local responders at this level. The Act describes the duties of appropriate stakeholders to cooperate in a LRF, and formal meetings and allocations of work to responsible stakeholders. It broadens the understanding of civil contingencies activity that now includes planning, preparation, response, recovery and protection, and requires Category One and Category Two responders within a given locality to coordinate and prepare for the causes and consequences of various events. The coordination, however, is event-specific and the participation of Category Two responders and other stakeholders depends on the type, location and scale of the event.

The LRFs, which are defined along Police Constabulary boundaries, typically meet three times a year to discuss emergency planning within their area and where relevant to liaise with neighbouring LRFs. In the event of a major emergency, the LRF would form a Strategic Coordinating Group for that emergency, that is, it would provide a forum for the co-ordination of a multi-agency response.

However, LRFs are neither a legal entity nor do they have powers to direct their members, which is often seen as a weakness of such a system, particularly when requiring the often voluntary involvement of private sector stakeholders such as utility companies. Therefore, while the Civil Contingencies Act has provided a useful framework for multi-stakeholder planning for emergency events, the decision to devolve powers to local public sector authorities and representatives of key private sector organisations is problematic. It effectively shifted the 'responsibility' (to take action) but not the 'blame' (if anything goes wrong). Another key problem is the emphasis on 'planning to respond' to emergency events rather than 'planning to reduce or eliminate' the emergency events.

Search terms: Local resilience forum; Civil Contingencies Act; emergency planning; resilience in the UK

8.3.6 Principle 6: Integrate Built Environment and Emergency Management Practitioners Into the DRR Process

The construction industry should adopt a disaster risk management perspective urgently. Professional institutions and trade associations should enhance the awareness of their members of their actual and potential roles in disaster risk reduction, as it will help to take necessary precautions at all stages of planning, design, construction and operation process. This is however difficult due to

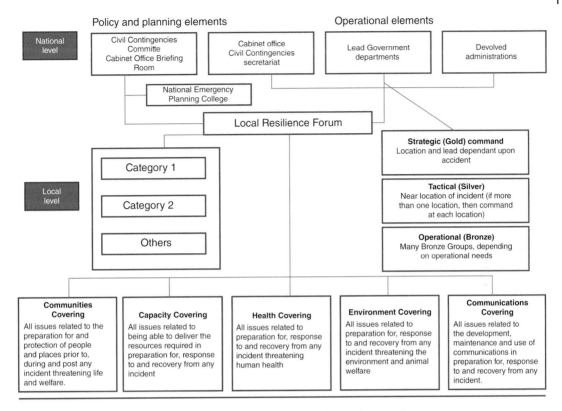

Figure 8.11 The Local Resilience Forum structure (*Source:* Adapted from Fisher, 2014).

the fragmented relationships between the various actors involved in construction projects. In addition, the need for integration goes beyond joining-up practitioners' activities: many of the non-technological problems appear because of the disciplinary boundaries with emergency planning, scientific community and policy makers. The role of architects is particularly interesting in this regard because of their ability to influence profoundly the design process through their interpretation of a brief and including the specification and configuration of materials. This requires the blending of subjectivity and objectivity throughout the design process, which places architects in an excellent position for coalescing other influential participants around them in support of the resilience effort.

A 'design charrette' (see Case Study 8.6) can be a useful collaborative approach that can aid the integration of various disciplinary perspectives in the context of a specific design requirement or project.

8.3.7 Principle 7: Mainstream Resilience into the Built Environment Curricula

The integration of construction professions with processes associated with disaster risk reduction is not taking place as rapidly as it should. For this to take place, there has to be a real and sustainable commitment to mainstreaming the need for a resilient built environment into the education programmes of those who are ultimately going to responsible for planning, designing, constructing, operating and maintaining it. Higher education and professional training play a major role in the

---

CASE STUDY 8.5

Rebuilding for the Future: Making Post-Sandy New York More Resilient and Sustainable

Nearly 70,000 buildings in New York ranging from beach bungalows to sea view offices sit within the 100-year floodplain. This can increase to 90,000 buildings in the next 10 years, equivalent to housing for more than 440,000 people. To protect this population, the city introduced 21 changes to the zoning and building codes, 16 of which have been adopted after Superstorm Sandy (see Case Study 4.2). In addition, there was a recommendation to not only make the city's buildings more robust against a similar event but also to make them more efficient (for another example see Case Study 9.2); it included a plan to redesign tall, exposed structures to better handle wind, to plant flood and wind resistant trees and to use cool surfaces to reduce summer heat.

In the storm's aftermath, there were calls for a single big fix, like building sea gates that would close off New York's harbour to rising water. However, such large engineering solutions would require great investment; in addition, there is no guarantee that they would protect New York in the future with the weather patterns changing continually and unpredictably due to the effects of climate change. The solutions being tried out are more widespread, and humbler, including stone revetments on Coney Island Creek to prevent 'backdoor' flooding, and solar-powered streetlights etc. The main idea behind these projects is that instead of building barriers, the city wants to build an ecosystem, in which it will gradually learn how to live with water. These plans are in stark contrast to some of the redevelopment along the New Jersey shoreline, where many of the affected properties have been allowed to rebuild without incorporating any significant DRR measures.

Search terms: Rebuilding New York after Sandy; Superstorm Sandy: resilient New York; sustainable New York

---

integration of disaster risk reduction principles, however, this is more problematic than it may appear at first glance as there is a need for practitioners to respect the local context within which disaster risk reduction takes place. This will require a change in attitudes towards construction development and knowledge of how to design and construct for hazard mitigation. In order to champion new and old community-based approaches, higher education techniques and on-the-job training to equip those responsible for engaging local communities with the skills necessary for buy-in and involvement in resilient solutions will be required. It is also important to ensure that science and traditional knowledge should be reconciled, as science can enable traditional knowledge systems to be easily understood by the practitioner, and traditional knowledge enables scientific concepts to be translated into modes of communication that are locally understood.

Key messages from recent 'Global Assessment Reports on DRR' published by the UNISDR and the work of the International Council for Building (CIB – *Conseil International du Bâtiment*) have highlighted that construction professionals are a key part of the solution for making the built environment more resilient (CIB 2016). Recent research acknowledges that disaster risk reduction issues need to be more explicitly addressed by practitioners, professional institutions and educators serving the construction sector, ensuring that disaster risk reduction is no longer seen as a niche skill but widely accepted as an important professional competency.

This raises implications for the core education and continued professional training of practitioners involved in the design, planning, construction, operation and maintenance of our increasingly urbanised world (Bosher *et al.*, 2015). Arguably some hazards, such as fires and earthquakes, can

---

CASE STUDY 8.6

Integrating Multidisciplinary Perspectives Through the Design Charrette

A design charrette (from the French word for 'chariot' or 'cart') has been described as a collaborative workshop in which a group of designers (sometimes with other groups, or laypeople) work together on a specific problem. A design charrettes can be used as a way to examine the organisational principles and awareness of resilient design.

An example of how such an approach can be useful included a project team (of engineers, architects, urban designers and quantity surveyors) working on tender documentation to redevelop a multi-use civic area in Greater London (see Figure 8.12). During the charrettes, a design scenario was tabled, with supporting documentation; the design actions of the project team was the focus for a set of assessments to analyse differences in process, actions, conflicts and resolutions. The assessments were undertaken in three sections, namely, 1) hazard and threat identification; 2) considering the implications of the hazards/threats and relevant vulnerabilities upon the development and 3) what courses of action are needed to deal with the identified hazards/threats. The 'wrap up' session that concluded the charrette included a discussion about any necessary design changes, in light of the identified hazards/threats, and judgements on revised project designs. The key lessons learnt from a number of these design charrettes have been:

- A number of key design oversights (related to hazard mitigation) have been identified that are likely to result in tangible changes to the project designs.
- It was very helpful to work on a live ongoing construction project but this was deemed as being most useful when undertaken in the design/tendering processes (i.e., at the earliest stage possible), see Figure 8.13.
- Awareness within some organisation's design teams of suitable guidance to mitigate hazards was limited and it was clear that this awareness needed to be improved.
- The piecemeal approach to urban regeneration was viewed as a constraint to the consideration of DRR; that is, multiple projects on one site can contribute to lack of 'joined up' thinking and the poor integration of potential 'solutions'.

---

be encompassed within building standards, such as Eurocodes, which form a core part of professional training. However, the extent to which other prominent hazards such as floods and coastal storms are being considered in the education of construction practitioners is highly questionable. Indeed, it is unclear to what extent these practitioners can address the increasingly complex demands of creating urban developments and infrastructure able to cope with 'multiple cascading hazards'.

Thus, professional institutions that support and accredit education provision for different disciplines within the construction sector should take the lead in educating students about their roles in disaster risk reduction. This would need the support of key professional institutions such as the Institution of Civil Engineers (ICE), American Society of Civil Engineers (ASCE), Royal Institute of British Architects (RIBA), Chartered Institute of Building (CIOB), Royal Institution of Chartered Surveyors (RICS) and the Royal Town Planning Institute (RTPI), to name just a few. As part of this, an open dialogue needs to be initiated with the professional institutions about the feasibility of including disaster risk reduction as an important competency though core undergraduate training, on-the-job practical training and/or continued professional development courses.

Figure 8.12 Inter-disciplinary working with a project team at a design charrette (*Source:* Lee Bosher).

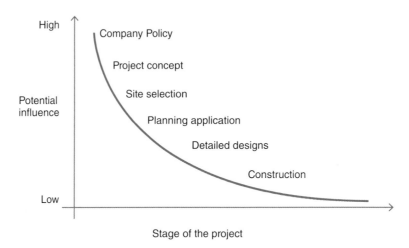

Figure 8.13 The 'Project influence curve' that illustrates that for concepts such as DRR to be made more influential, they need to be considered in the 'project concept' and ideally made a core component of 'Company Policy'.

8.4 Summary

A range of disasters in recent years have led to increased discussions about the way in which disasters can be avoided and managed. A paradigmatic shift has led to a focus on disaster preparedness, hazard mitigation and vulnerability reduction that simultaneously include both structural and non-structural approaches rather than the often reactive focus on disaster management and relief.

In addition, the role of a wider range of stakeholders – including construction professionals – in DRR has been emphasised, as it is important to enact the joined-up thinking necessary to mainstream DRR activities and thus resilience into the built environment. The Seven Guiding principles offer a point of departure when configuring methodologies for building in resilience in the future, and a framework for supporting the required shift towards more proactive disaster risk reduction. However, the real challenge is in developing the specific approaches necessary to account for individual contexts and situations. These may, of course, challenge some of the conventions that currently underpin construction development.

Accordingly, it is thus posited that by embedding DRR into core professional practice (including within company policy or at least the project concept as illustrated in Figure 8.13) it can be possible to make DRR part of the 'developmental DNA' and improve the future resilience of the built environment (Figure 8.14)

Key Points
- A holistic multi-hazard/threat and multi-stakeholder approach should be mainstreamed into development project in order to increase the resilience of the built environment.
- DRR measures should be considered and where relevant implemented at the earliest appropriate stage of the design and planning process rather than added onto a project as an after-thought.
- It is important to build back better, especially when opportunities arise in the aftermath of a disaster.
- Professional institutions can play a very positive role in educating students and members about their roles in disaster risk reduction.

Reinforcement
- What are the seven key principles for increasing the resilience of the built environment?
- What are the potential roles of construction stakeholders in the non-structural aspects of DRR?
- What participatory methods can be used to help generate inter-disciplinary design solutions to DRR?

Questions for discussion
- To what extent has a multi-stakeholder and multi-hazard approach to DRR become a basis for the Sendai Framework for Disaster Risk Reduction?
- Do you think it is actually appropriate to expect a broad range of construction professions to get more involved in DRR activities? If not, why not?
- Can you identify any examples (over and above the ones presented in this chapter) of how the seven guiding principles have been applied in the real world?

Further Reading

Books:

Bosher L.S., (ed.), (2008). *Hazards and the Built Environment: Attaining Built-in Resilience*, Taylor and Francis, London

Coaffee, J., Murkami-Wood, D and Rogers, P. (2008). *The Everyday Resilience of the City: How Cities Respond to Terrorism and Disaster*. Palgrave/Macmillian

Lizarralde G., Davidson C.H. and Johnson C., (eds.), (2010). *Rebuilding after disasters: From Emergency to Sustainability*, Spon Press, London

Mileti, D (1999), *Disasters by design*, USA: Joseph Henry Press

Vale L.J. and Campanella T.J., (eds), (2005), *The Resilient City: How Modern Cities Recover from Disaster*, Oxford University Press, Oxford

Articles:

Bosher, L. (2014). Built-in resilience through disaster risk reduction: Operational issues, *Building Research and Information*, 42(2), pp.240–254.

Bosher, L. and Dainty, A. (2011). 'Disaster risk reduction and 'built-in' resilience: towards overarching principles for construction practice', *Disasters*, 35(1), pp. 1–18

Bosher, L. Dainty, A., Carillo, P. and Glass, J. (2007). 'Built-in resilience to disasters: a pre-emptive approach', *Engineering, Construction and Architectural Management*, 14 (5), 434–446

Bosher L.S., Johnson C. & Von Meding J., (2015). 'Reducing disaster risk in cities: moving towards a new set of skills', *Proceedings of the ICE - Civil Engineering*, 168(3): pp. 99

Cartalis, C. (2014). Toward resilient cities – a review of definitions, challenges and prospects. *Advances in Building Energy Research*, 8(2), 259–66.

Carpenter, S., Walker, B., Anderies, J. M., and Abel, N. (2001). From metaphor to measurement: resilience of what to what? *Ecosystems*, 4(8), 765–781.

Chmutina, K., Ganor, T. and Bosher, L.,(2014). 'The role of urban design and planning in risk reduction: who should do what and when', *Proceedings of ICE – Urban Design and Planning.* 167(3), 125–135

Djalante, R. (2012). 'Adaptive governance and resilience: the role of multi-stakeholder platforms in disaster risk reduction', *Nat. Hazards Earth Syst. Sci.*, 12, 2923–2942

Hollnagel, E. (2014). 'Resilience engineering and the built environment', *Building Research & Information*, 42(2), 221–228

Lewis, J. (1999). *Development in Disaster-prone Places: Studies of Vulnerability*. Intermediate Technology Publications, London

Menoni, S. (2001). 'Chains of damages and failures in a metropolitan environment: some observations on the Kobe earthquake in 1995'. *Journal of Hazardous Materials*. 86(1–3). pp. 101–119

Mitchell, D., Enemark, S. and von Molen, P. (2015). 'Climate resilient urban development: Why responsible land governance is important', *Land Use Policy*, 48, 190-98

Shaw, K., and Maythorne, L. (2011). 'Managing for local resilience: towards a strategic approach', *Public Policy and Administration*, 28(1), 43–65

Reports:

CIB (2016), '*Disasters and the built environment: A Research Roadmap*', International Council for Research and Innovation in Building and Construction (CIB), Delft (Authored by Bosher L., Von Meding J., Johnson C., Farnaz Arefian F., Chmutina K. and Chang-Richards Y.) ISBN 978-908-0302-2-04

ISDR, 2008. *Indigenous knowledge for disaster risk reduction*. Available at: http://www.unisdr.org/we/inform/publications/3646

Kelman, I (2014). *Disaster mitigation is cost effective. Background note.* Available at: http://siteresources.worldbank.org/EXTNWDR2013/Resources/8258024-1352909193861/8936935-1356011448215/8986901-1380568255405/WDR14_bp_Disaster_Mitigation_is_Cost_Effective_Kelman.pdf

ODI (2015). *Unlocking the triple dividend of resilience*. Available at: https://www.gfdrr.org/sites/default/files/publication/unlocking_triple_dividend_resilience.pdf

9

DRR and Sustainability: An Integrated Approach

Development cannot be sustainable if the disaster risk reduction approach is not fully integrated into development planning and investments… Development investment that does not consider disaster risk will lead to the accumulation of more risk.

UN Secretary General Ban Ki-moon

Since the 1970s, people concerned with the world's so-called 'wicked problems' (i.e., interconnected issues like environmental degradation, poverty, food security and climate change) have marched together under the banner of 'sustainability'. The premise for the idea of sustainability is that with the right mixes of incentives, technology substitutions and social change, humanity can finally achieve a lasting equilibrium with the planet, and with one another. This idea has become even more popular in recent years, a period that has witnessed some of the hottest summers and coldest winters on record and an increasing array of extreme weather events. Amongst a growing number of scientists, community leaders, non-governmental organisations, governments and corporations, a new dialogue is emerging around the idea of resilience as a key component of sustainable development that includes long-term perspectives of how to help vulnerable people, organisations and systems persist, and possibly even thrive, amidst foreseeable and unforeseeable disruptions.

9.1   Learning Objectives

By the end of this chapter you will learn:

- What we mean by a sustainable built environment
- That to be truly sustainable the built environment should be also be resilient.
- What the opportunities and challenges are for creating a resilient and sustainable built environment?

9.2   Integrating Resilience and Sustainability: Why is it Important?

Superstorm Sandy (see Case Study 4.2) hit New York City particularly hard where it had most recently been redeveloped: Lower Manhattan. After the terrorist attacks of September 11, 2001, Lower Manhattan contained the largest collection of LEED-certified (see Box 9.1), green buildings in the world; this area it was rebuilt to be 'sustainable', not resilient. But arguably, and admittedly upon reflection, that was answering only part of problem. The buildings were designed to generate lower

Disaster Risk Reduction for the Built Environment, First Edition. Lee Bosher and Ksenia Chmutina.
© 2017 John Wiley & Sons Ltd. Published 2017 by John Wiley & Sons Ltd.

Figure 9.1 Sustainable houses at the University of Nottingham's Green Street (*Source:* Reproduced with permission of Lucelia Rodrigues).

Box 9.1 What Is LEED?

LEED ('Leadership in Energy & Environmental Design') is a green building certification program that recognizes best-in-class building strategies and practices. To receive LEED certification, building projects satisfy prerequisites and earn points to achieve different levels of certification.

---

environmental impacts, but not to respond to the impacts of the environment, for example, by some building having redundant power systems or storm resistant glazing (Figure 9.2).

Major disasters like Typhoon Haiyan, Superstorm Sandy (see Case Study 4.2), and earthquakes in Japan, Nepal and Haiti (see Case Study 5.2) were a wake-up call to remind built environment practitioners that it is critical to always keep the long-term picture in mind— and to think about DRR as early as possible during the design, construction and operation process, to prepare buildings to withstand possible hazards and threats as well as more commonplace longer-term maintenance issues required to reduce everyday wear and tear (see Thinking Point 9.1).

This, however, also has to be done with sustainability in mind. Achieving substantial cuts in Green House Gas (GHG) emissions is one of the most pressing challenges facing countries all around the world. One way to reach substantial emissions reduction is to widely adopt more sustainable forms of energy production accompanied by a reduction in energy consumption; thus, the challenges of restructuring the energy system and refurbishing the building stock have to be addressed. Rising electricity demand and fuel prices, decarbonisation of supply, liberalisation of markets and concern over energy security also encourage this transition.

Whilst being one of the largest CO_2 emissions contributors, buildings are vulnerable to the effects of climate change (as highlighted by recent flooding in the UK, see Case Study 3.1, and summertime

Figure 9.2 Blackout in Lower Manhattan after Superstorm Sandy (*Source:* Dan Nguyen on Flickr).

overheating). Where 'sustainability' aims to put the world back into balance, 'resilience' looks for ways to manage an imbalanced world. These issues emphasise the importance of energy and emissions reduction going hand in hand with resilience, without reducing the security, function and sustainability of the built environment.

THINKING POINT 9.1

Is Being Resilient Enough?

Resilience is often, rather myopically, understood as an ability to 'bounce back'. It is seen as a part of the climate change adaptation strategy: resilience helps an adjustment in natural and human systems, in response to actual or expected disturbances when frequencies tend to increase (e.g., building higher and higher levees in response to increasing risks of flooding). In addition, resilience incorporates coping at the local level (and sometimes even at the individual level) and dealing effectively with a single disturbance. It includes the understanding that a crisis is rare and temporary and that the situation will eventually normalise. But is coping enough when the ultimate goal is climate change mitigation?

When considered together, resilience and sustainability offer a transformation: the decisions made and actions taken change the identity of the system itself, and create a fundamentally new system when ecological, economic, and social structures make the existing system weak. Such synergy also addresses the causes of the increasing intensity and frequency of disturbances—that is, impacts of climate change. There are numerous examples of urban regions already engaged in developing both coping and adaptive strategies in response to, for example, sea level rise, demographic changes, and shortage of natural resources. However, with the increasing intensity and frequency of the disturbances, building, for example, larger dams or higher levees may no longer protect a city from flooding. Instead, a transformation to, for instance, a floating city that is self-reliant in terms of energy production, may be the only viable option for the long term.

9.3   What is Sustainability?

While the term sustainability in today's lexicon often conjures up an image of electric vehicles and wind turbines, 'sustainability' literally means 'to endure'. Any so-called 'green' products and buildings that don't stand the test of time are not truly sustainable. As climate change turns the attention to the possibility of increasingly likely disaster scenarios, it becomes clear that resilience should be discussed as part of the same conversation.

Sustainable development is not a single or well-defined concept; new meanings are continually added to this term. The idea of sustainable development goes back to 1970s. Its theoretical framework evolved after the publication of 'The Limits to Growth'. In the same year, the UN Conference on Human Environment was the first major international gathering to discuss sustainability on a global scale. It created a series of recommendations that later led to the establishment of the UN Environmental Programme (UNEP). The most famous definition was given in 1987 by the Brutland Commission, stating that sustainable development *'meets the needs of the present without compromising the ability of future generations to meet their own needs'*.

9.3.1   Understanding the Concept: Three Dimensions of Sustainability

Many argue that sustainability can be seen in three dimensions (Figure 9.3).

1) 'Techno-centric concerns', which encompass techno-economic systems, represent human skills and ingenuity (e.g., the skills that built environment practitioners must continue to deploy) and the economic system within which they are deployed.
2) 'Eco-centric concerns' represent the ability of the planet to sustain humans: both by providing material and energy resources and by accommodating us and our emissions and wastes.
3) 'Socio-centric concerns' represent human expectations and aspirations: the needs of human beings to live worthwhile lives, summed up by the phrase in the UK Government's interpretation of sustainable development as 'a better quality of life for everyone, now and in the future'.

Many, however, argue that culture and politics should also be included, and that the concept of sustainability should include more non-environmental aspects, with this change being driven by the

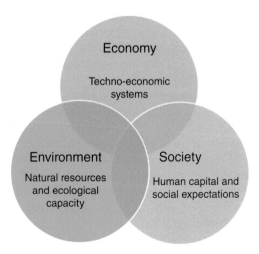

Figure 9.3 Three dimensions of sustainable development.

needs of the developing countries and is driven mainly by the discussion about the Millennium Development Goals and Sustainable Development Goals (see Case Study 9.1).

9.3.2 Global Challenges and Sustainability Index

The 15 Global Challenges provide a framework to assess the global and local prospects for humanity. Their description enriched with regional views and progress assessments are updated each year since 1996 and published in the annual State of the Future report. The Challenges are interdependent: an improvement in one makes it easier to address others; deterioration in one makes it harder to address others. Arguing whether one is more important than another is like arguing that the human nervous system is more important than the respiratory system. These Challenges are transnational in nature and trans-institutional in solution. They cannot be addressed by any government or institution

---

CASE STUDY 9.1

How Can DRR Help Achieve Sustainable Development Goals?

Disasters impact on every aspect of living targeted by SDG	Selected Sustainable Development Goals (SDG)	DRR to protect lives and the living, which is the key to SDG
In Aceh (Indonesia) the 2004 tsunami is estimated to have increased the proportion of people living below the poverty line from 30% to 50%.	1: End poverty in all forms everywhere; 2: End hunger, achieve food security and improved nutrition, and promote sustainable agriculture 11: Make cities and human settlements inclusive, safe, resilient and sustainable	Better land use planning enhances food productivity and strengthens sustainability.
The 2008 Sichuan earthquake destroyed 7,000 classrooms	4: Ensure inclusive and equitable quality education and promote lifelong learning opportunities for all	Only through building earthquake-proof schools can seismic prone countries and regions protect children and their education
61% of those that died in cyclone Nargis (Myanmar) were female	5: Promote gender equality and empower women and girls	Participatory DRR puts women at the forefront of protecting and sustaining their communities.
During the 2005 Pakistan earthquake, the estimated number of pregnant women in affected areas was 40,000.	3: Ensure healthy lives and promote well-being for all at all ages	Shelters built to protect communities against natural hazards can function as medical facilities and schools
There were over 17,000 cases of diarrheal disease after flooding in Bangladesh in 2004.	6: Ensure availability and sustainable management of water and sanitation for all	Water and sanitation systems built to hazard-resistant standards can resist becoming a breeding ground for disease.
Cyclone Nargis affected 16,800 ha of natural forest and 21,000 ha of forest plantations.	15: Protect, restore and promote sustainable use of terrestrial ecosystems, sustainably manage forests, combat desertification and halt and reverse land degradation, and halt biodiversity loss	Switching natural barriers such as mangroves in coastal areas and wooded hillsides can provide protection from cyclones and landslides/floods

Search terms: Millennium Development Goals; Disaster risk reduction; UN

acting alone, but require collaborative action among governments, international organizations, corporations, universities, NGOs, and creative individuals. The 15 Global Challenges are:

1) Sustainable development and climate change
2) Clean water
3) Population and resources
4) Democratisation
5) Global foresight and decision-making
6) Global convergence of IT
7) Rich–poor gap
8) Health issues
9) Education and learning
10) Peace and conflict
11) Status of women
12) Transnational organized crime
13) Energy
14) Science and technology
15) Global ethics

These global challenges are at the heart of the sustainability indicators, which are considered to be a good vehicle in helping to measure sustainable development and achieved progress. In 1995, the UN Division for Sustainable Development and the Statistics Division in close collaboration with international organisations and UN member states developed a set of 134 national indicators of sustainable development. These indicators are voluntary, and meant to assist member states in their work of reviewing their existing indicators or developing new indicators to measures progress towards nationally defined goals. The indicators have since been revised and decreased to 96 indicators, which are grouped into seven themes, and reflected in chapters of Agenda 21. In 2015, the State of the Future Index (SOFI) was introduced; it is an indication of the 10-year outlook for the future based on historical data of selected variables for the previous 20 or more years and on judgments about the best and worse plausible 10-year outcomes for each variable. It is constructed with key variables that are individually forecast and that in aggregate can indicate the potential trend of the future. The aim of these indicators is to be cross-cutting rather than represent four pillars (social, economic, environmental and institutional). The cross-cutting nature emphasises the multidimensional nature of sustainability, and these indicators are closely linked to Sustainable Development goals.

An overview of a SOFI is presented in Figures 9.4a and 9.4b, which shows where humanity is making progress (Figure 9.4a) and where more political attention and efforts are needed (Figure 9.4b). The world seems to be making progress in more areas than it is regressing or stagnating in. But since the areas of stagnation or regress are crucially important for the human and planetary survival, addressing them should be a top priority.

Box 9.2 Agenda 21

Agenda 21 is a non-binding, voluntarily implemented action plan of the United Nations with regard to sustainable development. It is a product of the Earth Summit (UN Conference on Environment and Development) held in Rio de Janeiro, Brazil, in 1992.

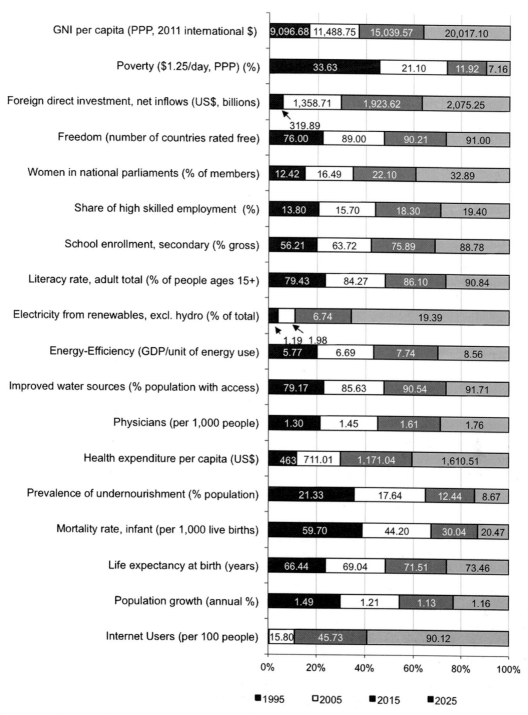

Figure 9.4a The State of the Future Index: where we are winning (*Source:* Reproduced with permission of The Millenium Project).

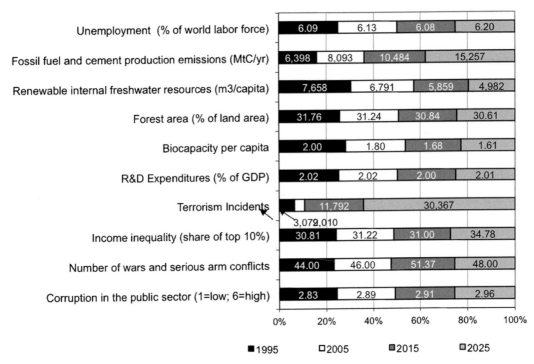

Figure 9.4b The State of the Future Index: where we are losing or there is no progress (*Source:* Reproduced with permission of The Millenium Project).

In many cases the relationship between indicators and policy is very strong with the policy framework in effect determining the indicators. While there may be concerns about having indicators closely aligned with policy and hence potentially biased towards particular policy priorities at the expense of other aspects of sustainable development, this is also one of their strengths. Policy makers see them as being directly relevant to the policies they have established and effective for communication. In several countries and institutions, the indicators are presented as an integral part of a sustainable development strategy, whether identified explicitly or generically. Commitments are made to report regularly on the indicators, and in some instances, commitments go as far as taking action if the indicators are not reporting favourable trends.

9.3.3 Sustainable Built Environment

Cities—and the built environment—are a large contributor to climate change. In highly developed countries, buildings account for 30–40% of total energy consumption, which is larger than industry sector (Figure 9.5). For example, in the United States, buildings account for 39% of CO_2 emissions, and in addition, they are the largest consumers of electricity and producers of waste. Buildings create a problem in low- and middle-income countries, too, mainly because of their high growth rate. By 2020, more than half of the building stock of China will have been constructed during the previous 15 years. All this demonstrates the importance that sustainable buildings could play in attaining 'sustainable development.'

There is not really an agreed definition of a sustainable built environment. There are many various assessment systems that rate whether the buildings are sustainable or not, and according to these systems, a building is sustainable if it is built in an ecologically oriented way, which reduced its impact over the environment. But is it enough? This definition is limited to the physical boundaries of the

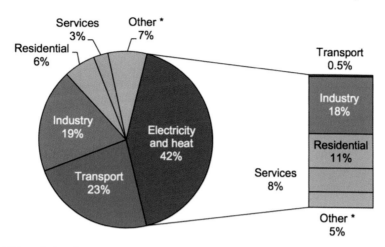

Figure 9.5 World CO_2 emissions by sector 2013 (*Source:* Reproduced with permission of IEA, 2014).

building and it mainly (or even only) sees sustainability through the environmental perspective. However, if taking a broader sustainability definition into account, sustainable built environment is a healthy facility designed and built in a cradle-to-grave resource-efficient manner, using ecological principles, social equality, and life-cycle quality value, which promotes a sense of sustainable community. Thus, this broader vision of a sustainable building suggests the need for an increase in:

- Demand for safe building, flexibility, market and economic value;
- Neutralisation of environmental impacts by including its context and its regeneration;
- Human well-being, occupants' satisfaction and stakeholders' rights;
- Social equity, aesthetics improvements, and preservation of cultural value.

9.4 Can the Built Environment Be Sustainable and Resilient?

A resilient system has the ability to bounce back from challenges, unharmed, and a big part of building in resilience may include building in ways that structural redundancy is incorporated, requiring the use of more resources (i.e., structural reinforcements or concrete) than would otherwise be needed. Sustainability, on the other hand, means efficiency, at least in part, as built environment practitioners strive to strike a balance between human needs and environmental impacts. Thus in theory, it appears that there can be tensions between achieving a resource efficient (green or sustainable) built environment or an intentionally resource inefficient (i.e., engineered redundancy and thus more resilient) built environment. To further complicate matters, over the next few decades the

---

Thinking Point 9.2

Is There a Difference Between a Green and a Sustainable Building?

Is a sustainable building the same as green building? These terms are often used interchangeable, but they are not really the same. The concept of green buildings appeared much earlier—in the mid-twentieth century when several communities driven by the ambitious of an ecological world advocated greet buildings. Green buildings were required to be disconnected from the service grids and made of natural materials. But in the 1970s, the turn to sustainability was taken due to the oil crises, and the buildings were then focused on for saving energy, which still remains the case.

Table 9.1 Comparing green and sustainable buildings.

Major areas of building performance	Green building	Sustainable building
Consumption of non-renewable resources	✓	✓
Water consumption	✓	✓
Material consumption	✓	✓
Land use	✓	✓
Impacts of site ecology	✓	✓
Urban and planning issues	(✓)	✓
GHG emissions	✓	✓
Solid waste and liquid effluents	✓	✓
Indoor well-being (air quality, lighting, acoustics)	(✓)	✓
Longevity, adaptability, flexibility		✓
Operations and maintenance		✓
Facilities management		✓
Social issues (access, education, inclusion, cohesion)		✓
Economic consideration		✓
Cultural perception and inspiration		✓

Sustainable buildings have a much broader meaning including more of a focus on economic and social requirements. It is becoming more common that the sustainability of the building is assessed by taking into account not only the operation stage of the building but also the materials, which are used as well as the whole life-cycle approach.

CASE STUDY 9.2

Kungsbrohuset Office Building, Stockholm

Kungsbrohuset is a 27,000 m^2 13-storey property in the centre of Stockholm, near the Stockholm Central Station (Figure 9.6); it has replaced the old, worn-out Kungsbrohuset building with a highly energy-efficient structure. It houses shops, restaurants, cafes, as well as a hotel and offices. The building has been built using tried and tested materials and smart solutions aimed at minimising the environmental impact.

The objective of the project is to create a development where the environment and energy efficiency are central considerations. An important part of the project is the building's energy-efficient facade, which controls the amount of sunlight allowed into the internal space and creates a balance so that the interior is neither too hot nor too cold. One of the most interesting features of the building is that every hour the Swedish meteorological centre automatically sends a detailed weather forecast direct to Kungsbrohuset via the GSM network. This means that the heat supply to Kungsbrohuset will be increased several hours before a drop in temperature occurs. By doing this, building's heat regulating system is put to optimal use.

The adjacent Lake Klara can be used for cooling, and the body heat of people passing on the street outside can be used for heating. Approximately 200,000 people pass through nearby Stockholm Central Station every day. These people, along with the various restaurants and cafés in the station, produce a large amount of excess heat. Instead of letting this heat goes to waste, it now being used for heating the Kungsbrohuset building. The technology is not new and is similar to the principles of Canadian wells. The heat generated by human activity is recovered by the ventilation system and then sent away from the station in its water tanks. Heated by this hot air, the water is sent to the second building through a heat exchanger, which allows the building to save up to 20% of its energy consumption. To complete the loop, the cooled water is returned to the station to optimize its temperature control. It must be said that the pipes were already in place and were long enough to extend into the adjacent building.

The building's branding is about being 'green': accordingly the office space is let to the companies that want to boost their image as environmentally friendly. Building users are offered coordinated framework agreements for cleaning services, servicing of office machinery and waste management. The building is also provided with secure bicycle storage area with a compressor pump and tools available for bicycle repairs, as well as dressing rooms and showers. The building is advertised as being 'eco-smart', which includes three characteristics:

- Eco-smart building: The building has an energy-efficient façade and environmentally adapted materials, combined with other innovative solutions that lead to three environmental certifications.
- Eco-everyday: Services and technical solutions that enable users to operate in an eco-friendly way.
- Eco-location: The building's proximity to public transport makes travelling and transports easier and contributes to lower CO_2 emissions.

Figure 9.6 Kungsbrohuset Office Building, Stockholm (*Source:* Ksenia Chmutina).

> **Box 9.3 Gigalopolis**
>
> Gigalopolis refers to the growing urban areas containing billions of people worldwide. Urban settlements and their connectivity will be the dominant driver of global change during the twenty-first century.

Figure 9.7 Urbanisation in Asia: Manila (*Source:* Ksenia Chmutina).

world's megacities will grow to become 'Gigalopolises' (see Box 9.3 and Figure 9.7). Efficiency in the construction of infrastructure will become a vital consideration because over the next few decades some countries will attempt to construct as much infrastructure that was built over the last few centuries.

Some of the most obvious ways to become more resilient are not sustainable. For example, the concerns about reliable electricity can be addressed by increasing the resilience of the local grid by buying a diesel generator (or two, or more). But extra diesel generators are certainly not an efficient, or particularly sustainable, way to create electricity. It's not ideal for the environment to be burning extra diesel, especially considering the knock-on impacts of air pollution.

It is therefore apparent that resilience and sustainability—two buzzwords within built environment disciplines—are terms that are discussed widely but used in an increasingly ambiguous way. In reality, and as already alluded to, these concepts are actually in tension with each other (see Thinking Point 9.3 and Figure 9.8).

9.4.1 Opportunities

However, deeper understanding of resilience and sustainability demonstrates that synergising these two ideas will present new interesting opportunities for ways the built environment can be developed, as it is clear that resilience can support sustainability greatly (Figure 9.9).

Thinking Point 9.3

Main Tensions Between Resilience and Sustainability

1) Achievements Versus Capacities: The sustainability agenda largely focuses on what can be obtained in terms of energy consumption performance and CO_2 emission reductions. This is simultaneously the cause and the consequence of an important emphasis on standards and performance assessment methods in the sustainability agenda. On the other hand, the resilience approach puts emphasis on what is available in order to cope with risks and threats.

2) Incremental Performances Versus Trial and Error Performances: Whereas sustainability calls for a maximisation of resources leading to the minimisation of resource consumption, resilience focuses on testing performances based on anticipated scenarios. Policy on sustainability has been based on incremental performances (arguably, affected by political interest). However, difficulties to reach these incremental performances have become obvious as not many countries are meeting their GHG reduction targets. Resilience policy, on the other hand, has largely focused on specific types of risk.

3) Efficiency Versus Redundancy: Sustainability suggests a lean approach to development and streamlining processes that reduce consumption and minimise environmental impacts. Resilience is significantly more open to assume the risk of overdesign in order to avoid damages and prevent disasters.

4) Mid-Term Equilibrium Versus Long-Term Equilibrium: There is no doubt that both sustainability and resilience call for a consideration of long-term impacts. However, they do it in a different way. Sustainability seeks that interventions made today do not affect what the next generations can do, based on targets that are defined today. On the other hand, resilience favours an ecological approach to regeneration that is not necessarily based on today's measures but in a constant reassessment of contingencies, resources, assets and risks.

5) Emphasis on Standards Versus Emphasis on Potentials: Sustainability has largely focused on globally accepted standards, often underestimating the importance of local conditions. Instead, resilience calls for the identification and reinforcement of local potentials of the system.

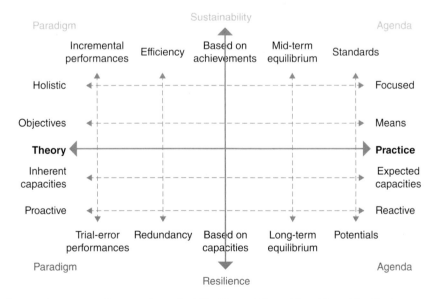

Figure 9.8 Tensions between resilience and sustainability (*Source:*Adapted from Lizarralde, 2015).

Resilience-building	Long-term **sustainable development**
Increasing societal resilience	Cohesive society with strong coping capabilities
Institutional capacity building	Strong, agile and prepared institutions
Physical reconstruction	Robust physical infrastructure and critical lifelines
Increasing environmental resilience	Environmental protection and adaptation strategies

Figure 9.9 Synergising resilience and sustainability.

A number of initiatives worldwide such as offshore windfarms located in Flevoland flooded areas (Figure 9.10) to protect coasts from flooding or security features that simultaneously act as sustainable urban mechanisms are becoming more and more common (see Case Study 9.3).

Other solutions may include:

- Pavements that can act as sustainable urban drainage systems (SUDS) and collapse if accessed by heavy vehicles: they can protect from localised/pluvial flood risk and the threat of vehicle-borne bombs; at the same time, SUDS can harvest rainwater that can then be used for functions that do not require treated water from the mains (flushing toilets, irrigation, cleaning paths/streets, etc.), which may contribute to water efficiency;
- Cool roofs reflecting the sunlight heatwave/overheating: they help keeping the space cooled down during the heatwaves if there is a blackout, and simultaneously reduces indoor temperature and helps reduce air-conditioning, and therefore, energy consumption (Figure 9.12);

Figure 9.10 An example of simultaneous use of resilient and sustainable measures in the Netherlands (*Source:* Reproduced with permission of Rogier van der Weijden).

CASE STUDY 9.3

New York Building Resiliency Task Force

In the aftermath of the Superstorm Sandy (see Case Study 4.2) 'The Building Resiliency Task Force Report' was issued; it provides 33 actionable proposals for making New York buildings and residents better prepared for the next extreme weather event taking into account that the weather events in the future could be more extreme due to the impacts of climate change. The Task Force provides recommendations on how to make sustainable and resilient various types of the building (Figure 9.11):

- The level of preparation for commercial buildings, both large and small, is fundamentally a business decision for their owners. Task Force recommendations are intended to minimise interruptions to building functionality while allowing the market to dictate the need to implement resiliency measures. Still, the city has an overall interest in maintaining a viable economy by reducing large-scale business disruption.
- Multifamily residences, dorms, hotels, and adult care facilities must provide for essential needs such as safety, drinking water, habitable temperatures, and functioning stairs and elevators. The Task Force intent was to add few financial burdens, and only in the most critical areas, given the limited financial resources available for upgrades.
- One- to three-family homes must have protection against storm damage and adequate emergency planning, as water can be supplied without pumps and vertical transportation is not an issue. Since many homeowners have limited financial resources for upgrades, the Task Force avoided adding significant financial burdens.

Although some of these recommendations are USA-focused, the majority can be adopted in any context, without requiring large additional investments.

Search terms: Superstorm Sandy; rebuilding New York; Building Resiliency Task Force; resilient New York; sustainable New York; post-Sandy New York.

- Energy co-generation system: incorporation of the off-site renewable energy technology can provide electricity during a blackout as a result of damage to the electricity infrastructure as well as contribute to the overall decrease of energy from fossil fuels;
- Trees around the building: they can stop vehicles from coming close to a building thus decreasing security threats, and at the same time can provide shading thus keeping the space from overheating, which leads to the reduction in overall energy consumption;
- Use of window shutters: protects glazing from being damaged during storms and hurricanes as well as protecting spaces from overheating thus reducing energy consumption needs associated with the use of air-conditioning (see Figure 9.13);
- Anti-blast glazing incorporating photovoltaic (solar cells) film: anti-blast glazing can reduce the impact of a terrorist attack or strong windstorms whilst generating energy thus reducing the need for energy consumption from fossil fuels.

9.5 Summary

As architect Carl Elefante once said, "The greenest building is the one that's already built," so the goal of all built environment practitioners should be to design, construct and operate a built environment

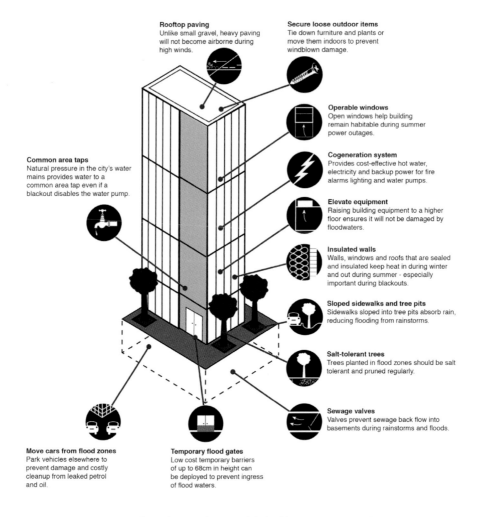

Rooftop paving
Unlike small gravel, heavy paving will not become airborne during high winds.

Secure loose outdoor items
Tie down furniture and plants or move them indoors to prevent windblown damage.

Operable windows
Open windows help building remain habitable during summer power outages.

Common area taps
Natural pressure in the city's water mains provides water to a common area tap even if a blackout disables the water pump.

Cogeneration system
Provides cost-effective hot water, electricity and backup power for fire alarms lighting and water pumps.

Elevate equipment
Raising building equipment to a higher floor ensures it will not be damaged by floodwaters.

Insulated walls
Walls, windows and roofs that are sealed and insulated keep heat in during winter and out during summer - especially important during blackouts.

Sloped sidewalks and tree pits
Sidewalks sloped into tree pits absorb rain, reducing flooding from rainstorms.

Salt-tolerant trees
Trees planted in flood zones should be salt tolerant and pruned regularly.

Sewage valves
Valves prevent sewage back flow into basements during rainstorms and floods.

Move cars from flood zones
Park vehicles elsewhere to prevent damage and costly cleanup from leaked petrol and oil.

Temporary flood gates
Low cost temporary barriers of up to 68cm in height can be deployed to prevent ingress of flood waters.

Figure 9.11 Example components of a resilient and sustainable building (*Source:* WEDC).

that will last for a long time without putting extra pressures on humans and the natural environment. Photovoltaics and low-flow toilets are definitely moves in the right direction, but such measures are not enough to fully attain 'sustainability'; the built environment needs to be able to stand the test of time.

Key Points:
- Though the variables, which contribute to resilience, are many and often complicated, buildings need to be resilient in order to be truly sustainable.
- A resilient and sustainable built environment is a complex and multi-faceted idea that requires long-term thinking about worst-case disaster scenarios as well as more common, everyday wear as part of the future performance measures.

Reinforcement:
- What are the main differences between sustainability and resilience?

Figure 9.12 Example of a cool roof on a BASF House at the University of Nottingham's Green Street (*Source:* Reproduced with permission of Luceila Rodrigues).

- What are the main drivers for synergising the ideas of resilience and sustainability in the built environment?

Questions for discussion:
- What other sustainable and resilience measures can you think of that can be introduced in the modern built environment?
- What are the main challenges that slow down the simultaneous implementation of the resilience and sustainability in the built environment practice?

Figure 9.13 Typical 'chattel house' with window shutters; vernacular housing found in Barbados (*Source:* Lee Bosher).

Further Reading

Books:

Drexler, H. and Khouli, S. (2012). *Sustainable by Design: Methods for Holistic Housing, Basics, Strategies, Projects.* Basel, Swizerland: Birkhauser.

Haas, T (ed.) (2012). *Sustainable Urbanism and Beyond: Rethinking Cities for the Future.* NY: Rizzoli International Publications.

Shaw, R., Pulhin, J.M., and Pereira, J. J. (eds) (2010). *Climate Change Adaptation and Disaster Risk Reduction: Issues and Challenges.* Vol. 4. Emerald.

Sinclair, C and Sothr, K. (2006). *Design Like You Give a Damn: Architectural Responses to Humanitarian Crises.* NY: Metropolis Books.

Articles/reports:

Berardi, U. (2012). 'Sustainability Assessment in the Construction Sector: Rating Systems and Rated Buildings', *Sustainable Development*, 20(6), 411–424

Berardi, U. (2013). 'Clarifying the new interpretations of the concept of sustainable building', *Sustainable Cities and Society*, 8, 72–78

Brundtland, G. H. (1987), *Report of the World Commission on environment and development: Our common future*, New York: United Nations

Coaffee, J. (2008). 'Risk, resilience, and environmentally sustainable cities', *Energy Policy*, 36 (12) 4633–4638

Coaffee, J. and Bosher, L.S., (2008). 'Integrating counter-terrorist resilience into sustainability', *Proceedings of ICE Urban Design and Planning*, 161 (DP2), 75–83

Dainty, A., & Bosher, L. (2008). 'Integrating resilience into construction practice': In Bosher L.S., (Ed), *Hazards and the Built Environment.* New York: Routledge, 357–372

Korhonen, J., and Seager, T. P. (2008). 'Beyond eco-efficiency: A resilience perspective', *Business Strategy and the Environment*, 17(7), 411–419

Lizarralde, G., Chmutina, K., Dainty, A. and Bosher, L., (2015). 'Tensions and complexities in creating a sustainable and resilient built environment: Achieving a turquoise agenda in the UK', *Sustainable Cities and Society*, Vol. 15, pp. 96–104

Perelman, L. J. (2008). 'Infrastructure risk and renewal: The clash of blue and green'. Paper presented at the 4-18 Jan., *Public Entity Risk institute Symposium.* Fairfax, Virginia, USA

Redman, C.I. (2014). 'Should sustainability and resilience be combined or remain distinct pursuits?', *Ecology and Society*, 19(2)

Thornbush, M., Golubchikiv, O., & Bouzarovski, S. (2013). 'Sustainable cities targeted by combined mitigation–adaptation efforts for future-proofing', *Sustainable Cities and Society*, 9(1–9)

Tobin, G. A. (1999). 'Sustainability and community resilience: the holy grail of hazards planning?', *Global Environmental Change Part B: Environmental Hazards*, 1(1), 13–25

Reports:

ARUP (2014). *Cities alive: rethinking green infrastructure.* Available at: http://www.arup.com/ Homepage_Cities_Alive.aspx

SwissRe (2014). *Economics of Climate Adaptation –Shaping climate-resilient development.* Available at: http://media.swissre.com/documents/rethinking_shaping_climate_resilent_development_en.pdf

UN (2015). *Global sustainable development report.* Available at: https://sustainabledevelopment.un.org/ content/documents/1758GSDR%202015%20Advance%20Unedited%20Version.pdf

10

Conclusions and Recommendations

Opening image: The urban sprawl of Iran's capital of Tehran, stretching out northwards towards the tectonically active Alborz Mountains (*Source:* Lee Bosher, 2009).

Disaster Risk Reduction for the Built Environment, First Edition. Lee Bosher and Ksenia Chmutina.
© 2017 John Wiley & Sons Ltd. Published 2017 by John Wiley & Sons Ltd.

While hazards, such as earthquakes, floods, storms and tsunamis, are natural in origin, the way that disaster risk has been created and increasingly embedded into the contemporary urban landscape is largely human induced. Decades of mass urbanisation accompanied by expanding socio-economic inequalities, poor urban planning, non-existent or poorly regulated building codes and little or no proactive adaptation to the impacts of climate change has increased humanity's exposure to these hazards.

10.1 Dynamic Factors (and Root Causes)

The last century has witnessed an expansion of urban populations across the world. This mass urbanisation has occurred in the context of neo-liberal 'free-market' policies in which the role of the state as an urban custodian has gradually been diluted (Johnson *et al.*, 2013). For urban planning and construction, this has resulted in a reduction in regulatory control and a perspective that the role of the state is primarily to enable 'free' markets to work. The implication for the construction sector is to enable investments in construction through the provision of infrastructure, financial mechanisms and making land available for development. However, reduced (or ineffectively applied) regulatory controls have meant that disaster risks, and other environmental concerns, are often poorly considered in urban development decisions (UNISDR, 2011; Johnson *et al.*, 2013).

Clearly, there are a myriad highly ingrained root causes of why some people/regions/nations are poor/marginalised and/or particularly prone to the impacts of natural hazards. These root causes, such as global and regional political ideologies, power relations, trade agreements and regulations are suitably articulated in publications such as 'At Risk' by Wisner *et al.* (2004) and thus are not discussed here in depth, but where relevant they are alluded to. With a focus on what Wisner *et al.* (2004) might term 'dynamic pressures', Table 10.1 provides a summary of some of the key dynamic factors that invariably contribute towards turning natural hazards into large-scale disasters (and typically not 'natural disasters'!).

10.2 Moving away from Disaster Risk Creation

The relative increase over the last two decades in the number of large-scale disasters, the amount of people being affected and the economic losses generated (UNISDR, 2013a) indicates this situation needs to improve sooner rather than later. It is even argued by experts such as Ben Wisner, that rather than overly focusing on DRR we should also consider, to at least the same extent, the concept of 'disaster risk creation' as a function of business-as-usual economic, social and political transactions. Wisner raises a very good point and thus key stakeholders such as planners and commercial developers clearly need to be a central part of these considerations.

It has been highlighted by the UN's 'Global Assessment Report' in 2015 (UNISDR, 2015), that Disaster Risk Reduction activities need to move away from an approach constrained within disciplinary siloes dominated by emergency management and civil protection practitioners. Consequently, it has been argued in this book, that while these practitioners still have an important role to play, the broad range of people responsible for the delivery, operation and maintenance of the built environment also need to become more proactively involved in making the built environment resilient to a wide range of hazards and threats. Accordingly, this book has attempted to highlight the (actual and potential) roles of a wide range of stakeholders associated with the integration of disaster risk reduction into the (re)development of the built environment.

Table 10.1 Overview of Some of the Main Factors Contributing Towards Disasters.

Hazard type	Key dynamic factors contributing towards disaster risk creation
Floods	• Poor (or unregulated) urban planning that results in homes, offices, infrastructure and essential services being built on flood-prone land. • Buildings and infrastructure not designed to cope with the physical impacts of floods. • Increased urbanisation leading to more rapid run-off from surfaces to water courses • Lack of, or ineffective, urban drainage systems that struggle to cope with expanding developments or the impacts of climate change. • Poorly conceived river/flood management schemes leading to an over reliance of physical assets that need long-term maintenance and can become quickly outdated. • Deforestation of steep slopes leading to instable surfaces, increased run-off and siltation of water courses. • The construction of dams leading to flooded tracts of land; some dams can also pose flood risks due to the potential structural failures. • Lack of, or ineffective, emergency preparedness and evacuation procedures.
Windstorms	• Buildings and infrastructure not designed to cope with the physical impacts of windstorms • Poorly conceived coastal/flood management schemes leading to over reliance of physical assets that need long-term maintenance and become quickly outdated. • Long-term draining of water logged coastal areas that then become protected from the deposition of alluvial soils resulting in land that can subside below safe levels and prone to storm surges. • Deforestation of natural coastal vegetation and forests that can act as natural buffers providing some protection from the impacts of strong winds and storm surges • Poor (or unregulated) urban planning of homes, offices, infrastructure and essential services on land prone to the impacts of storms and storm surges. • Lack of, or ineffective, emergency preparedness and evacuation procedures.
Earthquakes	• Poor (or unregulated) urban planning resulting in homes, offices, infrastructure and essential services being built in seismically active locations, particularly locations prone to liquefaction. • Lack of reliable (uncorrupted) code enforcement mechanisms • Buildings and infrastructure not designed to cope with the physical impacts of earthquakes. • Artificial development of land converted from wetland areas or the sea; such land can be particularly prone to the impacts of liquefaction. • Lack of, or ineffective, emergency preparedness and evacuation procedures.
Volcanoes	• A long established urge for humans to utilise the very fertile soils that typically lie close to volcanoes • Poor, or unregulated, land use planning that results in homes, offices, infrastructure and essential services being built in locations prone to the impacts of ground related volcanic hazards (such as lahars, lava flows, pyroclastic flows and landslides). • Lack of, or ineffective, emergency preparedness and evacuation procedures.
Landslides	• Expansion of settlements and infrastructure into land increasingly exposed to slope instabilities • Deforestation of steep slopes leading to instable surfaces and increased run-off. • Increased urbanisation leading to more rapid run-off from surfaces to water courses • Development of land on (or near to) steep slopes that have not been protected (sufficiently) from slope instabilities. • Impacts of climate change leading to increased extreme precipitation in areas prone to landslides. • Lack of, or ineffective, emergency preparedness and evacuation procedures.

There is a need to embrace DRR in a way that enables individuals and society to adapt (pro-actively) to not only micro-level issues but also macro level issues (such as climate change). This implies a need to adapt the way citizens live their lives (at the individual, household, community, town, regional, country levels). This adaptation will need to incorporate the physical, natural, institutional, economic, educational, community/social and individual components of DRR. With particular relevance to new developments and coming back to the specific roles of built environment professionals, a range of strategies are required to help attain improved social, physical and institutional resilience to a diversity of hazards and threats. These measures are likely to include:

Utilising internationally recognised risk management frameworks – The international risk management standard ISO 31000 'Risk management – Principles and guidelines' (British Standards Institution, 2009) has proven to be very useful as a common framework for risk management across sectors. ISO31000 presents four stages, those being risk identification, assessment, evaluation, and treatment that are the backbone to the risk management framework used throughout this book. Since the framework is based upon an accepted international standard such as ISO 31000, it is anticipated that it will provide suitable relevance (in functionality and terminology used) globally. It is intended that this straightforward risk management approach (summarised in Figure 10.1) can be applicable to a range of urban contexts (i.e., city, building, organisation); a broad range of professions (i.e., planners, architects, security consultants, engineers, local authorities); and covering a range of countries (Pan-European and global).

Embracing a multi-hazard/threat approach to understanding disaster risk – It is important to recognise that in this book natural hazards have been presented in a rather compartmentalised way (i.e., geological hazards described as separate from hydro-meteorological hazards). In reality it is likely that these hazards will coexist, not only with other hazards but also with a complex mix of human induced threats (malicious or non-malicious). Embedded in the risk management approach endorsed in this book is the requirement to undertake multi-hazard/threat risk assessments that can help to identify the most prominent issues as well as whether there are opportunities to development risk reduction approaches that address the risk posed by more than one hazard. Conversely, there is a need to appreciate that some risk reduction measures may be suitable for dealing with one type of hazard but not appropriate for other types of hazards or threats (for instance, some measures for dealing with riverine flood risk may not be suitable for addressing seismic risk). Consequently, multi-hazard/threat assessments should be undertaken and any risk reduction options should be proportionately considered alongside other hazards or threats that have been identified.

Revisions to building codes – It is likely that any required revisions to building codes will necessitate investment in generating and utilising more research based evidence that encompasses a range of structural and non-structural approaches to addressing multiple hazards. Of course the governance mechanisms (including financial support) need to be provided so that appropriate building codes are also enforced by qualified professionals.

Tightening of planning policy - A strong political will is required (at all levels of governance) to think and act for the long-term (i.e., 25-100 year time scales) and move away from the 'Not In My Term of Office' (NIMTOO) attitude. Central to this is the need to address institutional failures at multiple scales, looking especially at the role of corruption as well as management strategies such as public private partnerships and outsourcing of regulatory and other government functions.

Encouraging a socio-technical systems approach - Good practice guidance is needed on the broad range of structural and non-structural risk reduction measures. Suitable information also needs to be specifically designed for the general public; such as people that buy or rent houses and the extent to which they are aware of the hazards that may affect their 'investments'. As part of this, the

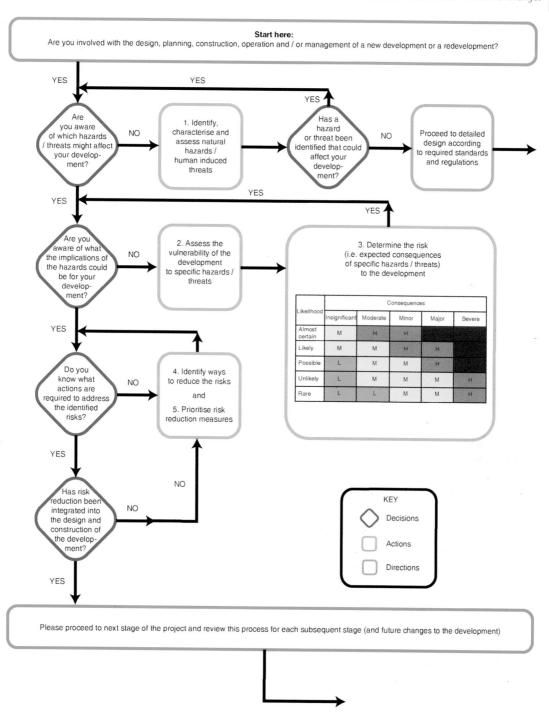

Figure 10.1 Overview of the risk management decision making framework.

way that information is provided to the general public could be improved and made more accessible. There is a need to open up political and social space within which local residents can assess their own vulnerability to the hazards to which they give priority and develop capabilities to reduce risk while pursuing other goals.

Improvements to professional training - As advocated by Russell (2013) and Janda and Pareg (2013) new skills are required as core competencies to enable a better understanding of the societal aspects of construction professions and improved engagement/empowerment with stakeholders (such as clients and local communities). Bosher *et al.* (2015) take this idea further and state that the professional institutions that provide education to, and accredit courses for, construction professionals should take the lead in educating students about their roles in disaster risk reduction. Table 10.2 provides a sample list of some of the key institutions that could be involved in such an initiative. It is, therefore, important that the wide range of suitably trained built environment professionals (as detailed in Chapter 8) are consulted and actively involved in the strategies that will be required to make built assets more resilient to natural hazards and human induced threats. While admittedly this is not a panacea it would definitely be a move in the right direction.

Table 10.2 A Non-Exhaustive List of Some of the Key Professional Institutions that Could Support the Integration of DRR Training Into Courses for a Range of Construction Professionals.

Professional Institution	Acronym
American Society of Civil Engineers	ASCE
American Society for Engineering Education	ASEE
American Society of Safety Engineers	ASSE
Chartered Association of Building Engineers	CABE
Chartered Institute of Building Service Engineers	CIBSE
Chartered Institute of Architectural Technology	CIAT
Chartered Institute of Building	CIOB
Chartered Institution of Highways and Transportation	CIHT
Chartered Institution of Water and Environmental Management	CIWEM
European Network for Accreditation of Engineering Education	ENAEE
Engineering Council	
Institute of Highway Engineers	IHE
Institution of Civil Engineers	ICE
Institution of Professional Engineers New Zealand	IPENZ
Institution of Structural Engineers	IStructE
Japan Society of Civil Engineers	JSCE
Royal Institute of British Architects	RIBA
Royal Institution of Chartered Surveyors	RICS
Royal Town Planning Institute	RTPI
UK Standard for Professional Engineering Competence	UKSPEC

10.3   Moving towards a New Developmental DNA

Although it may be impossible to accurately predict everything (in the present, let alone the future), it is just as important to recognise that disasters often occur because risk reduction measures have not been considered or undertaken, despite there being prior knowledge of existing hazards and threats. DRR approaches may sometimes be limited because they primarily address the 'foreseeable' events. However, such approaches may engender adaptive capabilities that can prove useful when dealing with any future unforeseen (natural or human-induced) events. When societies become more effective at dealing with the known or anticipated risks, it is expected that this may stimulate a 'culture of resilience and adaptability' that can enable them to deal with some of the less foreseeable events.

The idea of societies being able to generate a culture that can adapt to survive internal and external threats is not new. Indeed there are relevant lessons from previous European societies. Gerrard and Petley (2013) claim that evidence exists to support the view that European societies in the Middle Ages (AD 1000–1550) were quite advanced in developing sophisticated measures (structural and non-structural) to cope with natural hazards such as floods and earthquakes. If this was the case, have modern (more urbanised) societies lost these capabilities to adapt?

The Sendai Framework's predecessor, the Hyogo Framework for Action (HFA) was adopted in 2005 not long after the Indian Ocean tsunami killed approximately 227,000 people. Over 120 countries participated in the last HFA reporting cycle (2013–2015) detailing their progress in reducing disaster risk. This regular voluntary monitoring of implementation of these agreements is evidence that a culture of disaster risk reduction is gradually spreading across the world and explains why many countries are experiencing some success in reducing mortality from disasters, particularly climate-related disasters which now account for 90% of all disasters linked to natural hazards.

Some of the operational issues highlighted earlier in this book indicate that the private sector has a critical role to play in proactively addressing DRR. The realities of free-markets economics (i.e., that places a value on flood-prone land and a competitive market for insurers to provide flood insurance as standard) and the lack of incentives for the private (and even the public-private) sector to proactively consider DRR on developments have resulted in a legacy of inappropriately conceived developments. These developmental practices have occurred to promote economic development, but not necessarily to enable appropriate development.

Nonetheless, Bosher (2014) believes that there is scope for utilising an approach to DRR that is less dependent on governmental regulation. For instance, possibly through forward thinking private sector developers that can grasp the business opportunity (even if it is just driven by free-market fundamentalism). For some 'new build' developments, particular developers are recognising that it could actually be a good idea to become a market leader in incorporating DRR into commercial developments with the hope that it will give them the cutting edge over competitors in the short term (i.e., under risk-blind legislative conditions) and the long term (see UNISDR, 2013b for some specific examples).

This brings to mind the 'Project influence curve' from Chapter 8 (Figure 8.13) that illustrated for DRR ideologies to be made more influential, they need to be considered in the 'project concept' and maybe even made a core component of 'Company Policy'. Figure 10.2 moves this idea on a stage further and embraces Allan Lavell's suggestion that we need to change the way we develop our cities, infrastructure and buildings, by not merely mainstreaming DRR into practice but by making DRR part of the 'developmental DNA'. Accordingly, Figure 10.2 illustrates how by considering DRR within 'company policy' and the 'project concept' it should be possible to make DRR part of the developmental DNA. If DRR is only considered in the planning and detailed design stages then there is hope that DRR measures will be included but they may not be highly effective. If DRR is not considered or

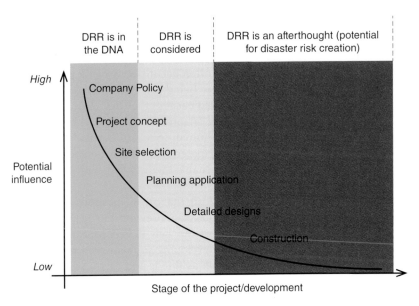

Figure 10.2 The 'Project influence curve' illustrates that by making DRR a core component of 'Company Policy' or 'Project concept' it can be possible to make DRR part of the developmental DNA.

only considered once construction or reconstruction has started then the creation of disaster risk is much more likely.

Of course, this will be deemed overly optimistic by detractors of the free-market mantra but now is arguably not the time to give up on optimism. Additionally, this will all need to be supported by many other non-structural activities such as, and not limited to, incorporating DRR into the professional training (formal and informal) of built environment professionals and raising awareness of proactive risk reduction to deal with the current and longer-term impacts of climate change.

10.4 Future Research and Educational Challenges

It is argued that if adopted holistically, DRR could be an effective way to attain a more resilient built environment (see Twigg, 2007; Bosher and Dainty, 2011). Encouraging a wide range of stakeholders to proactively integrate DRR considerations into the (re)development of the built environment is thus frequently being promoted but in reality is often stymied by a range of operational issues that have raised a number of research challenges as summarised in Table 10.3. Of particular importance is the need to develop (or adapt existing) urban multi-hazard/threat risk assessment methodologies so that the potential 'spin-off benefits' of dealing with different hazards can be maximised (i.e., flood risk management approaches that also provide benefits to increased security and/or sustainability) and any clashes can be minimised.

10.5 Final Thoughts for Construction Practitioners

During the last few decades a paradigmatic shift has contributed towards an increased focus on disaster preparedness, hazard mitigation and vulnerability reduction rather than the often reactive focus on disaster management and relief. Despite this new emphasis, the construction industry at various scales is arguably poorly positioned to embrace the tenets of DRR. The construction industry's structural fragmentation sustained by ingrained practices which have emerged from the temporal nature of projects arguably present a problematic arena within which to enact the joined-up thinking

Table 10.3 Overview of the Key Operational Issues and Relevant Research and Educational Challenges (adapted from Bosher, 2014).

Operational issues	Research and educational challenges in relation to disaster risk
1) Addressing the legacy of inappropriate urbanisation	• Develop relevant urban multi-hazard/threat risk assessment methodologies • Develop (and understand the cost/benefits of) innovative new technologies/materials and/or improve the existing technologies/materials for reducing risks and enhancing protection and retrofitting • Better understand how existing properties can be made more resilient/resistant to hazards • Learn lessons from how other countries cope with and indeed live with disaster risk and ensure these lessons are used to in the planning and design of new and existing urban developments • Assess most effective (from functional and cost perspectives) ways to repair damaged properties so that they will perform better the next time an hazardous event occurs
2) Attending to the impacts of climate change	• Make accurate climate change data and scenarios available to a broader range of professions (i.e., architects, civil engineers, urban planners) • Better understand the impacts of climate change on modern methods of construction • Compile progressive interactive data banks on a range of natural hazards so that risks from one hazard can be understood/mitigated in the context of other hazards • More strongly understand the conceptual and operational links between the resilience and sustainability agendas, what are the clashes and synergies? • Establish an open access database of good practices for DRR in different contexts (i.e., different hazards/threats, urban contexts, pre-/post-disaster, types of construction professionals).
3) Legislation only goes part of the way	• Develop (and if possible adopt relevant existing) multidisciplinary frameworks for the application of mainstreamed DRR into the process of planning and design, implementing and construction, management and maintenance of the built environment in both pre-disaster and post disaster situations. • Learn lessons from how neighbouring nations cope with and indeed live with the natural hazards (particularly flood risk) and ensure these lessons are used to in the planning and design of new and existing urban developments • Put more power in the hands of local communities, possibly by strengthening the advocacy roles they can play in influencing the planning process and reducing 'hazard-blind' developments
4) Increasing the awareness of who should do what	• Educate the broad range of disciplines responsible for the delivery and operation of the built environment so that they know about the notion of DRR in relation to their specific areas of specialism. This can be achieved through Continued Professional Development (CPD) type courses but also, ideally, incorporated into the professional training (i.e., undergraduate/apprentice) of engineers, architects and planners. • Produce bespoke guidance briefs for specific disciplines highlighting their roles in reducing/eliminating disaster risk on developments/re-developments

(Continued)

Table 10.3 (Continued)

Operational issues	Research and educational challenges in relation to disaster risk
5) Improving when key decisions should be made	• Educate the broad range of disciplines responsible for the delivery and operation of the built environment so that they know about the notion of DRR in relation to their specific areas of specialism. • Eliminate/reduce planning loopholes by strengthening planning systems so that disaster risk creation is avoided not ignored • Produce bespoke guidance briefs for specific disciplines highlighting their roles in reducing/eliminating disaster risk on developments/re-developments • Provide training to SMEs in the construction sector on how to resiliently repair disaster affected properties
6) Understanding who pays and the 'business case'	• Better understand the costs and benefits of DRR approaches with the aim of supporting an evidence-based 'business case' for proactive DRR. • Understand how commercial demand can be generated for new 'resilient' technologies so that they can become affordable to the private sector/public sector organisations (including SMEs). • Better understand the most cost-effective ways to resiliently reinstate disaster affected properties and encourage the insurance sector to adopt the best approaches.

necessary to mainstream DRR (Bosher and Dainty, 2011), let alone the more ambitious aim of DRR becoming part of the 'developmental DNA'.

The perspectives that have contributed towards the development of the seven guiding principles (after Bosher and Dainty, 2011) that were discussed in Chapter 8 highlight the varying levels of input that are required; embracing the formally and informally trained construction practitioners and artisans as well as the 'client' and the 'community', the governmental policy maker and the non-governmental organisation; in fact a significant range of decision makers that should be involved in the delivery of the built environment. The range of risk reduction options discussed in the preceding chapters highlight that non-structural as well as structural adaptations need to be considered to reduce the threat, and impact, of disasters and that lessons can be learnt from a range of disciplines and socio-cultural contexts. However, for this to be achieved governments, the commercial sector and civil society will urgently need to prioritise DRR as a critical development challenge and make concerted efforts to develop related policies, capabilities and legislative and institutional arrangements.

In attempting to join up and synthesise the various perspectives discussed in this book, it has been necessary to problematise the notion of DRR. Two factors stand out in this regard. Firstly, the complexity and interrelatedness of the array of continually evolving risks facing the built environment and its users present a continually shifting set of parameters against which built assets should be evaluated. This suggests a need for fresh criteria for establishing the appropriateness of construction development which embraces DRR. Secondly, the institutional and commercial resistance to change which pervades the construction industry stands as a considerable barrier to enacting the broad range of DRR approaches advocated within this book. Thus, although this book offers a point of departure for embedding DRR considerations in the future, and a framework for supporting the required shift towards more proactive DRR, the real obstacle is in challenging some of the conventions which currently underpin construction and urban development.

10.5.1 Towards DRR as a Core Professional Competency

It is apparent that the broad range of construction professionals need to do a better job at transferring existing knowledge; many of the problems being encountered in hazard prone developments are not about knowledge/information not existing (i.e., technical information on how to build flood resistant structures), it is primarily about the knowledge not being applied (for instance, due to poor knowledge transfer, poor training, commercial self-interests or poor regulation). Thus, there is a need for broadening the core skills base (the breadth of multi-hazard DRR considerations, rather than just specialising in specifics such as earthquake or wind engineering) so that non-structural approaches to DRR can be given as much credence as some of the more technical structural considerations.

It is thus argued by Bosher *et al.* (2015), and reiterated in these pages, that proactively dealing with disaster risk should not merely be a 'bolt on' consideration otherwise it tends to be more expensive, poorly integrated and less effective than if incorporated into earlier designs. This raises implications for the core education and continued professional training of the construction practitioners that are involved in the design, planning, construction, operation and maintenance of our increasingly urbanised world. Consequently, it is increasingly being argued that the institutions that provide built environment related education/training programmes should take the lead in educating students about their roles in DRR. This would need the support of key professional institutions (such as the ICE, RIBA, CIOB and RICS) including an open dialogue about the feasibility of including DRR as a professional competency though core undergraduate training, on-the-job practical training and/or Continued Professional Development courses. In the meantime it is hoped that the contents of the book will inspire a broad range of construction practitioners to incorporate DRR thinking and innovations into their everyday practice.

Further Reading

Books

British Standards Institution, (2009). *Risk management: Principles and guidelines*, London: British Standards Institution Group

UNISDR, (2013b). *Global Assessment Report 2013 on Disaster Risk Reduction*, UNISDR, Geneva, Switzerland

UNISDR, (2012). *Concept note: Fourth session of the Global Platform for Disaster Risk Reduction*, United Nations International Strategy for Disaster Reduction Secretariat, Geneva

UNISDR, (2011). *Global Assessment Report on Disaster Risk Reduction: Revealing Risk, Redefining Development.* United Nations International Strategy for Disaster Reduction, Geneva

Wisner B., Blaikie P., Cannon T., and Davis I. (2004). *At Risk: Natural Hazards, People's Vulnerability, and Disasters: Second Edition*, London, Routledge

Journals

Bosher L.S., (2014), 'Built-in resilience' through Disaster Risk Reduction: Operational issues', *Building Research & Information*, Vol. 42, No. 2, pp. 240–254

Bosher L.S. and Dainty A.R.J., (2011). 'Disaster risk reduction and 'built-in' resilience: Towards overarching principles for construction practice', *Disasters: The Journal of Disaster Studies*, Policy and Management, Vol. 35, No. 1, pp. 1–18

Bosher L.S., Johnson C. & Von Meding J., (2015). 'Reducing disaster risk in cities: moving towards a new set of skills', *Proceedings of the ICE - Civil Engineering*, 168(3): pp. 99

Gerrard C. M. and Petley D. N., (2013). 'A risk society? Environmental hazards, risk and resilience in the later Middle Ages in Europe', *Natural Hazards*, 69: 1051–1079

Janda, K.B. and Parag, Y. (2013). A middle-out approach for improving energy performance in buildings, *Building Research & Information*, 41(1), 39–50

Johnson C., Bosher L., Adekalan I., Jabeen H., Kataria S., Wijitbusaba A. and Zerjav B., (2013). 'Private sector investment decisions in building and construction: increasing, managing and transferring risks'. Working paper for the *Global Assessment Report 2013 on Disaster Risk Reduction*, UNISDR, Geneva, Switzerland

Russell, J. S. (2013). 'Shaping the future of the civil engineering profession', *Journal of Construction Engineering and Management*, 139(6): 654–664

Internet:

Twigg, J. (2009). *Characteristics of a disaster-resilient community*. Available at: http://practicalaction.org/docs/ia1/characteristics-disaster-resilient-community-v2.pdf

UNISDR (2015). *Global assessment report on disaster risk reduction* 2015, available at: http://www.preventionweb.net/english/hyogo/gar/2015/en/gar-pdf/GAR2015_EN.pdf

UNISDR, (2013a). 'Disaster statistics', *United Nations International Strategy for Disaster Reduction Secretariat*, Geneva Available on-line http://www.unisdr.org/we/inform/disaster-statistics (accessed 12th February 2013)

Index

Disaster Risk Reduction for the Built Environment, First Edition. Lee Bosher and Ksenia Chmutina.
© 2017 John Wiley & Sons Ltd. Published 2017 by John Wiley & Sons Ltd.